Chemical Fixation of Carbon Dioxide

Methods for Recycling CO_2 into Useful Products

Chemical Fixation of

Carbon Dioxide

Methods for Recycling CO_2 into Useful Products

Martin M. Halmann, Ph. D.

Department of Environmental Science and Energy Research
Weizmann Institute of Science
Rehovot, Israel

CRC Press

Boca Raton   Ann Arbor   London   Tokyo

Library of Congress Cataloging-in-Publication Data

Halmann, Martin.
    Chemical fixation of carbon dioxide: methods for recycling CO_2
    into useful products / Martin Halmann.
        p.   cm.
    Includes bibliographical references and index.
    ISBN 0-8493-4428-X
    1. Carbon dioxide—Recycling. 2. Reduction (Chemistry)
    I. Title.
    TP244.C1H35   1993
    665.8'9—dc20   92-35370
                                                                              CIP

Direct all inquiries to CRC Press, Inc., 2000 Corporate Blvd., N.W., Boca Raton, Florida 33431.

© 1993 by CRC Press, Inc.

International Standard Book Number 0-8493-4428-X

Library of Congress Card Number 92-35370

Printed in the United States of America   1 2 3 4 5 6 7 8 9 0

Printed on acid-free paper

Preface

Global warming, and the role of carbon dioxide in the greenhouse effect, are now widely recognized as urgent environmental problems. The usual solution has been to suggest decreases in fossil fuel consumption; the only major viable alternative would be increased usage of nuclear energy.

An alternative approach may be the chemical fixation of carbon dioxide. This is currently carried out on a large industrial scale in the production of urea from carbon dioxide and ammonia. Additionally, it is utilized in the synthesis of methanol and other petrochemicals from carbon dioxide, carbon monoxide, and hydrogen. In many countries numerous research efforts have been devoted, during the last 20 years, to developing additional tools for the recycling of carbon dioxide, by chemical reactions, into useful products. The methods used have been catalytic, photochemical, electrochemical, photoelectrochemical, and photocatalytic. Some of these reactions may be considered biomimetic, those which stimulate natural photosynthesis. Several of these reactions have been achieved with high efficiency and may be close to commercial applications. Others are still in an early stage of research, but may suggest interesting avenues for further development.

The purpose of this book is to bring together in one volume the remarkable results already obtained by the different methods. This book should be particularly helpful to researchers and students planning to embark on novel projects in environmental chemistry and to those who intend to benefit from the application of carbon dioxide as an inexpensive and widely available raw material for organic synthesis. Also, this book will be useful as an auxiliary text in advanced undergraduate or introductory graduate courses in environmental science.

The Author

Professor Martin Mordehai Halmann has been Professor Emeritus at the Weizmann Institute of Science, Rehovot, since 1990. He has been a member of the scientific staff since 1949. He studied chemistry at Hebrew University, Jerusalem, receiving his M.Sc. degree in 1949 and Ph.D. in 1952.

His main research interests are chemical reaction mechanisms, isotope effects on chemical reactions, isotope effects on vibrational spectra and on Franck-Condon factors, chemical effects of nuclear transformations, photochemistry of organic phosphorus compounds, photo-electrochemical and photocatalytic reduction of carbon dioxide using semiconductors, electrochemical and photochemical reactions in molten salts, photochemical oxidation of bromide to bromine in aqueous brines, and photosensitized oxidation of organic pollutants in water.

To my wife, Mirjam,
and my children, Michal and Nahi,
with love

Table of Contents

ONE

Introduction: Carbon Dioxide and Global Warming

Natural green plant photosynthesis is the most important chemical reaction on the Earth — and the basis of all life processes. Most of the Earth's carbon, 10^{16} tons, is deposited as carbonates, while about 10^{14} tons are estimated to be the total amount of carbon dioxide in the atmosphere and the oceans and thus available for photosynthesis.[1] From this, about 2×10^{11} tons of biomass are produced yearly by natural plant photosynthesis.[2] One of the largest carbon reservoirs on the Earth's surface is oceanic dissolved organic carbon (DOC). This is composed mainly of compounds such as humic substances, which are resistant to biological degradation, but which are photochemically degraded by sunlight to biologically labile compounds. Since the future depletion of atmospheric ozone will increase the solar UV flux, this photochemical degradation is likely to enhance the recycling of biologically refractory DOC.[3] On the other hand, the rate of growth of forests and of other biomass increases with the partial pressure of carbon dioxide. At present, the *fertilization effect* of increased carbon dioxide is considered to out-weigh the effects of forest clearance. Thus, the biota provides a large enough sink of atmospheric carbon dioxide to enable a balance of the global carbon budget.[4] One of the uncertainties is the amount of interhemispheric transport of carbon dioxide provided by ocean circulation.[5] The total uptake of anthropogenic CO_2 by the Earth's oceans has been estimated to be about 1 gigatons carbon (2×10^9 tons) per year.[6]

Among the components of air, carbon dioxide comprises only about 0.034% by volume. This carbon dioxide content is responsible for the acidity (pH 5.7) of precipitations from the troposphere, the natural *acid rain*.[7]

1

The ever-increasing consumption of fossil fuel since the industrial revolution some 150 years ago — together with specific industries such as cement production and, possibly, also the rapid deforestation of the Earth's tropical rain forests[8,9] — have caused a very marked increase in the global concentration of carbon dioxide, which now amounts to 7×10^9 tons/year. The annual increase (%) of the important greenhouse gases is presented in Figure 1.[10]

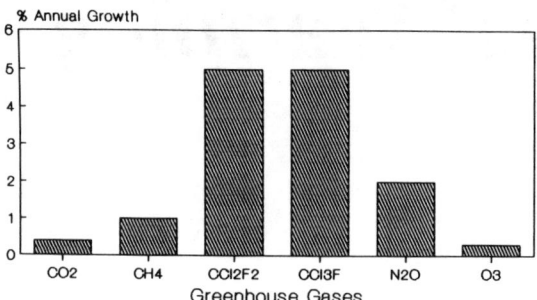

Fig. 1   Annual increase (%) in greenhouse gases.

Thus, from 1958 to 1980, the atmospheric carbon dioxide concentration increased from 315 to 340 ppm. It has been predicted that, if both fossil fuel combustion and forest clearance will continue at the present rate, atmospheric carbon dioxide concentrations will reach levels of 1000 to 2000 ppm within the next few centuries.[4]

This infrared-absorbing gas, together with other naturally produced gases, such as methane and nitrous oxide, threatens to cause an increase in the Earth's temperature, thus leading to the *Greenhouse Effect*. The warming contribution of the major greenhouse gases is shown in Figure 2.[10]

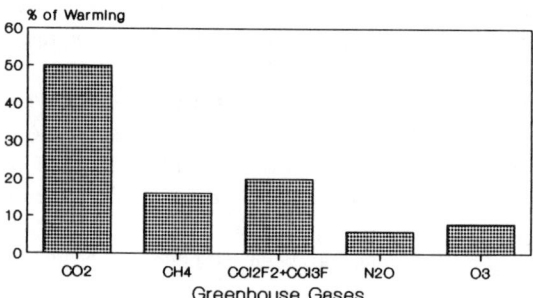

Fig. 2   Warming contribution by greenhouse gases.

Of the 50% contribution of carbon dioxide to the current global warming, about 35% is due directly to the effects of fossil fuel fired energy generation, while deforestation, agriculture, and industry are estimated to contribute 10, 3, and 2%, respectively.[10]

Efforts to decrease the consumption of fossil fuel are limited by rising human population and increasing industrialization, particularly in developing countries. Alternatives to fossil fuel all have important limitations.[11] For wider use of nuclear energy, the problems of reactor safety and of the safe disposal and storage of spent fuel elements have yet to be solved. Also, the limited amount of proven reserves of uranium requires the development and safe operation of breeder reactors. Nuclear fusion, in spite of massive research efforts, still seems far away from realization and cannot, at present, be considered a viable alternative energy source. Solar, wind, hydroelectric, and geothermal sources of energy, while important and worthwhile to develop, may probably not provide more than 10 to 20% of global energy requirements. In addition, fossil organic substances such as oil, coal, natural gas and, oil shales are also essential sources of organic raw materials, vital for the production of plastics, medicines, manmade fibers, agrochemicals, and many other valuable products.

It should thus be of considerable urgency not only to conserve as much as possible the known reserves of fossil fuel resources, but also to search for means to recycle their main degradation or combustion product, which is carbon dioxide. We shall consider the processes now used industrially for the chemical fixation of CO_2[12] (Figure 3) and which reactions provide a promise for the future.

Major Industrial Uses

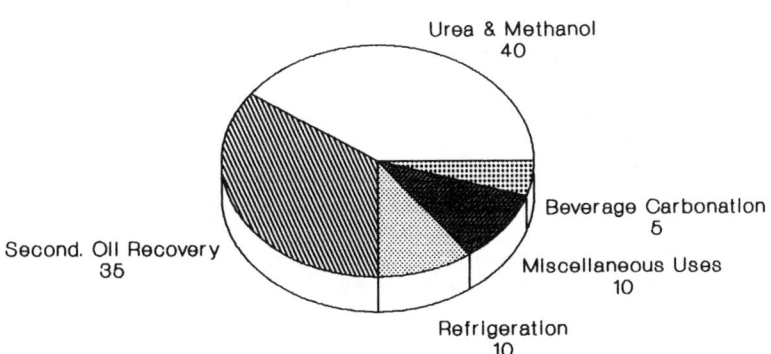

Fig. 3 Uses (%) of recovered carbon dioxide.

Several specialized reviews have described different aspects of the chemical fixation of carbon dioxide.[1,2,13-16]

References

1. **Behr, A.**, Use of carbon dioxide in industrial organic synthesis, *Chem. Eng. Technol.*, 10, 16–27, 1987.

2. **Behr, A.**, Carbon dioxide as an alternative C_1 synthetic unit: activation by transition-metal complexes, *Angew. Chem. Int. Ed. Engl.*, 27, 661–678, 1988.

3. **Mopper, K., Zhou, X., Kieber, R. J., Kieber, D. J., Sikorski, R. J., and Jones, R. D.**, Photochemical degradation of organic carbon and its impact on the oceanic carbon cycle, *Nature*, 353, 60–62, 1991.

4. **Walker, J. C. G. and Kasting, J. F.**, Effects of fuel and forest conservation on future levels of atmospheric carbon dioxide, *Palaeogeography, Palaeoclimatology, Palaeoecology (Global Planetary Change Section)*, 97, 151–189, 1992.

5. **Broecker, W. S. and Peng, T.-H.**, Interhemispheric transport of carbon dioxide by ocean circulation, *Nature*, 356, 587–589, 1992.

6. **Sarmiento, J. L. and Sundquist, E. T.**, Revised budget for the oceanic uptake of anthropogenic carbon dioxide, *Nature*, 356, 589–593, 1992.

7. **Wisseroth, K.**, Problem of atmospheric carbon dioxide and its possible control by an ocean system, *Chemiker-Zeitung*, 115, 45–52, 1991.

8. **Lugo, A. E. and Brown, S.**, Steady state terrestrial ecosystems and the global carbon cycle, *Vegetatio*, 68, 83–90, 1986.

9. **Brown, S., Gillespie, A. J. R., and Lugo, A. E.**, Biomass estimation methods for tropical forests with applications to forest inventory data, *Forest Sci.*, 35, 881–902, 1989.

10. **Mintzer, I. M.**, Energy, greenhouse gases, and climate change, in *Annual Reviews of Energy*, Vol. 15, Hollander, J. M., Socolow, R. H., and Sternlicht, D., Eds., 1990, 513–538.

11. **Dostrovsky, I.**, *Energy and the Missing Resource*, Cambridge University Press, 1988.

12. **Ballou, W. R.**, Carbon Dioxide, in *Kirk-Othmer Encyclopedia of Chemical Technology*, 3rd ed., Vol. 4, Wiley-Interscience, New York, 1978, 725–742.

13. **Halmann, M.**, Photochemical fixation of carbon dioxide, in *Energy Resources through Photochemistry and Catalysis*, Grätzel, M., Ed., Academic Press, New York, 1983, 507–534.

14. **Aresta, M. and Forti, G., Eds.**, *Carbon Dioxide as a Source of Carbon. Biochemical and Chemical Use*, Kluwer Academic, Hague, 1987.

15. **Braunstein, P., Matt, D., and Nobel, D.**, Reactions of carbon dioxide with carbon-carbon bond formation catalyzed by transition metal complexes, *Chem. Rev.*, 88, 747–764, 1988.

16. **Collin, J. P. and Sauvage, J. P.**, Electrochemical reduction of carbon dioxide mediated by molecular catalysis, *Coord. Chem. Rev.*, 93, 245–268, 1989.

TWO

Thermodynamics and Thermal Dissociation

THERMODYNAMICS

The equilibrium in the general reaction

$$\text{organic compounds} + O_2 = CO_2 + H_2O + \text{energy}$$

is at normal temperatures very much to the right, toward oxidation of organic compounds. Living plants absorbing sunlight reverse the direction of this reaction, by the natural process of photosynthesis. In artificial photosynthesis, the primary reactions of CO_2 reduction, with their free energy change per electron $E°$ are represented by the equations

$$CO_2 + H_2O(1) = HCOOH(aq) + 1/2O_2 \quad E° = 1.428 \text{ eV} \qquad \textbf{(1)}$$

$$CO_2 + H_2O(1) = HCHO(aq) + O_2 \quad E° = 1.350 \text{ eV} \qquad \textbf{(2)}$$

$$CO_2 + 2H_2O(1) = CH_3OH(aq) + 3/2O_2 \quad E° = 1.119 \text{ eV} \qquad \textbf{(3)}$$

$$CO_2 + 2H_2O(1) = CH_4(g) + 2O_2 \quad E° = 1.037 \text{ eV} \qquad \textbf{(4)}$$

$$CO_2 \rightarrow CO(g) + 1/2O_2 \quad E° = 1.33 \text{ eV} \qquad \textbf{(5)}$$

These reactions of carbon dioxide and water are assumed to lead directly to formic acid, formaldehyde, methanol, and methane, involving two, four, six, and eight electron transfers, respectively.[1]

The energy input in these reactions may be electrical or photonic, or both. The effects of elevated temperatures on these equilibria are such that the reduction of CO_2 directly to formaldehyde, methanol, or methane is favored by rising temperatures.[1]

THERMAL DISSOCIATION

Direct dissociation of carbon dioxide is possible only at very high temperatures,[2,3]

$$CO_2 = CO + 0.5O_2 \; \Delta G = 0 \text{ at } 3350 \text{ K} \qquad (6)$$

Such temperatures are accessible with solar furnaces, but the separation of the dissociated products presents formidable problems. A promising approach to the separation of the oxygen produced from the carbon dioxide and carbon monoxide is that using an oxygen semipermeable membrane, such as with calcia-stabilized zirconia in a solid-phase electrolyte[4] (see also Chapter 7).

On clean metal surfaces, carbon dioxide molecules adsorb in their linear un-reactive form, while on oxide surfaces they react to produce stable carbonates. The adsorption and reactions of CO_2 on clean Rh, Pd, Pt, Ni, Fe, Cu, Re, Al, Mg, and Ag metal surfaces were studied by spectroscopic methods, particularly with respect to the formation of CO_2^-, which, depending on the metal, may undergo dissociation to CO and O, or transform into CO_3 and CO. The binding energy of adsorbed CO_2 was increased by alkali adatoms. If oxygen was preadsorbed on the metals, the formation of stable carbonate structures was favored. On Pt and Cu metal surfaces, there was mainly molecular adsorption, with hardly any dissociation of CO_2. On the other hand, CO_2 adsorption was dissociative on Fe, Ni, Re, Al, and Mg surfaces. With catalyst-support systems of electron donating character, carbon dioxide may be activated by its conversion to a bent structure. This activation of carbon dioxide, leading to the adsorbed CO_2^- species, was achieved on platinum group metals by the use of preadsorbed alkali promoters.[5] With low coverage of alkali adatoms, such as with potassium of up to about 0.25 monolayer, the alkali atoms donate their 4s electrons to the metal surface, resulting in increase both of the rates of adsorption and of the dissociation of CO_2. On the other hand, at about monolayer coverage, carbon dioxide adsorbs in a stable form, presumably of a $K\text{-}CO_2$ surface complex. At low potassium coverage on Rh(III) surfaces, carbon dioxide was found to dissociate to CO and O already at 131 to 200 K. At high potassium coverage, the $K\text{-}CO_2$ formed was found to be trans-formed above 200 K into an adsorbed carbonate-like species,

$$2CO_2(ads)^- \rightarrow CO(ads) + CO_3(ads)^{2-} \qquad (7)$$

which decomposed to carbon dioxide only above 650 K.[6]

On clean rhodium surfaces such as on the Rh(111) and Rh(100) faces, even at low temperature (300°K), CO_2 dissociated to surface adsorbed CO and O,

$$CO_2(g) \rightarrow CO_{ads} + O_{ads} \qquad (8)$$

The dissociation probably occurred at surface defect sites, and was inhibited by oxygen pre-adsorption.[7] At 1 atm pressure and 500 K, CO_2 did dissociate rapidly.[8]

The adsorption of both carbon dioxide and ammonia at clean Cu(100) and Zn(0001) surfaces was investigated using a combination of X-ray photoelectron and electron energy loss spectroscopies (XPS and HREELS). While CO_2 and NH_3 separately, when adsorbed on either the Cu or Zn surfaces at low temperatures (80 K), show the electron energy loss spectra of these species (vibrational bands due to the stretching and bending modes of these molecules, and also some overtone bands), when heated above about 200 K, both CO_2 and NH_3 were completely desorbed from both metal surfaces. However, when CO_2 and NH_3 were coadsorbed at the Cu(100) and Zn(0001) surfaces at 80 K, and then warmed to 298 K, there appeared in the electron energy loss spectra new bands. These could be assigned to the vibrational bands of a surface-bound ammonium carbamate, NH_2CO_2. This species must have been formed by breaking an N-H bond. Thus, carbon dioxide was activated by its coadsorption with ammonia[9] (see also in Chapter 9 on the photoreduction of CO_2 with pre-adsorbed NH_3 over CeO_2-TiO_2).

Adsorption of CO_2 on polycrystalline TiO_2 (anatase form) and on Pt/TiO_2 was studied by Auger electron spectroscopy (AES) and X-ray photoelectron spectroscopy (XPS). The species identified on TiO_2 were graphitic carbon, HCO_3^-, CO_3^{2-}, adsorbed CO_2, TiC, as well as adsorbed CO on Pt. The adsorbed CO_2 molecules dissociated on Pt/TiO_2 to form CO molecules and an oxygen atom. No CO_2 dissociation occurred on unsupported Pt.[10]

THERMOCHEMICAL CYCLE

A variety of thermochemical cycles have been proposed, in which the thermal energy of solar furnaces or nuclear reactors may be used to supply the energy required for the splitting of water. Analogous thermochemical cycles may be useful for the thermochemical splitting of carbon dioxide, to achieve the overall reaction,

$$CO_2 + heat \rightarrow CO + 1/2O_2 \qquad (9)$$

One such cycle which has been tested depended on the reaction of cerium(IV) oxide with solid or molten sodium pyrophosphate at 750 to 950°C to form sodium cerium(III) phosphate, trisodium phosphate, and oxygen,[11]

$$2CeO_2(s) + 3Na_4P_2O_{7(s)} \rightarrow 2Na_3Ce(PO_4)_{2(s)} + 2Na_3PO_{4(s)} + 1/2O_{2(g)} \qquad (10)$$

The rate of oxygen evolution in this reaction step could be significantly enhanced by adding Li_3PO_4 to the reaction mixture, thus forming a lower-temperature melting eutectic composition, in the temperature range of 780 to 880°C. With a molar ratio $Li_3PO_4/Na_4P_2O_7 = 0.20$, the rate of O_2 production was four times larger that in the absence of the lithium salt. In the second reaction step, also at a high temperature (750 to 900°C), the sodium cerium(III) phosphate reacted with sodium carbonate to recover cerium(IV) oxide, and to produce carbon monoxide,

$$2Na_3Ce(PO_4)_{2(s)} + 3Na_2CO_{3(l)} \rightarrow$$
$$2CeO_{2(s)} + 4Na_3PO_{4(s)} + 2CO_{2(g)} + CO_{(g)} \tag{11}$$

In a variant of this reaction, steam was added in the above reaction mixture, thus promoting the water gas shift reaction, with formation of hydrogen and recovery of the carbon dioxide,

$$CO + H_2O \rightarrow CO_2 + H_2 \tag{12}$$

In a low-temperature recovery step (at 5 to 50°C), trisodium phosphate was treated with carbon dioxide in aqueous solution to yield disodium hydrogen phosphate and sodium bicarbonate,

$$6Na_3PO_4 + 6CO_2 + 6H_2O \rightarrow 6Na_2HPO_4 + 6NaHCO_3 \tag{13}$$

which could be separated by crystallization. These two compounds by moderate heating ($>250°C$ and $>200°C$) could be dehydrated to $Na_4P_2O_7$ and Na_2CO_3, respectively, thus closing the cycle. In contrast to most other thermochemical cycles, the above process avoids corrosive and toxic reaction intermediates. Also, the chemicals involved in this cycle need not be of high purity, and are thus relatively inexpensive.[11]

References

1. **Halmann, M.,** Photochemical fixation of carbon dioxide, in *Energy Resources through Photochemistry and Catalysis,* Grätzel, M., Ed., Academic Press, New York, 1983, 507–534.

2. **Martin, L. R.,** Use of solar energy to reduce carbon dioxide, *Solar Energy,* 24, 271–277, 1980.

3. **Lietzke, M. H. and Mullins, C.,** The thermal decomposition of carbon dioxide, *J. Inorg. Nucl. Chem.,* 43, 1769–1771, 1981.

4. **Nigara, Y. and Cales, B.,** Production of carbon monoxide by direct thermal splitting of carbon dioxide at high temperature, *Bull. Chem. Soc. Jpn.,* 59, 1997–2002, 1986.

5. **Solymosi, F.,** The bonding structure and reactions of CO_2 adsorbed on clean and promoted metal surfaces, *J. Mol. Catal.,* 65, 337–358, 1991.

6. **Solymosi, F.,** Adsorption, bonding and reactivity of CO_2 on clean and promoted metal surfaces, *Proc. Int. Symp. Chemical Fixation of Carbon Dioxide,* Nagoya, Japan, Dec. 2–4, 1991, 227–236.

7. **Dubois, L. H. and Somorjai, G. A.,** Comments on "Why CO_2 does not dissociate on Rh at low temperature" by W. H. Weinberg, *Surf. Sci. Lett.,* 128, 231, 1983.

8. **Weinberg, W. H.,** Why CO_2 does not dissociate on Rh at low temperature, *Surf. Sci. Lett.,* 128, 224–228, 1983.

9. **Davies, P. R. and Roberts, M. W.**, Activation of carbon dioxide by ammonia at Cu(100) and Zn(0001) surfaces leading to the formation of a surface carbamate, *J. Chem. Soc. Faraday Trans.*, 88, 361–368, 1992.

10. **Tanaka, K., Miyahara, M., and Toyoshima, I.**, Adsorption of CO_2 on TiO_2 and Pt/ TiO_2 studied by X-ray photoelectron spectroscopy and Auger electron spectroscopy, *J. Phys. Chem.*, 88, 3504–3508, 1984.

11. **Bamberger, C. E. and Robinson, P. R.**, Thermochemical splitting of water and carbon dioxide with cerium compounds, *Inorg. Chim. Acta*, 42, 133–137, 1980.

THREE

Dynamics of Carbon Dioxide Reduction

The reduction of carbon dioxide to useful compounds such as fuels and chemical intermediates is inherently difficult, as it involves several steps of both electron and proton transfers. The primary steps in the reactions of carbon dioxide or the bicarbonate or carbonate ions in solution were identified mainly by pulse radiolysis, but also by flash photolysis, electron spin resonance, and electrochemical experiments. The radiolysis of water results in short-lived radicals, ions, and atoms, which may react with certain solutes to produce secondary radicals. If these secondary radicals have intense UV- or visible absorption bands, the rates of their formation and decay can be determined by kinetic spectrophotometry. Their interaction with other solutes may result in increases in the decay rates, from which the second-order rate constants of such interaction can be derived.[1]

Electron capture by carbon dioxide in aqueous solutions results in formation of the $\cdot CO_2^-$ radical anion,

$$CO_2 + e^-(aq) \rightarrow \cdot CO_2^- \tag{1}$$

which strongly absorbs in the ultraviolet region, with $\lambda_{max} = 235$ nm and $\epsilon = 3000$ M^{-1} cm^{-1}.[2] The rate of reaction of the hydrated electron with carbon dioxide is close to the diffusion-controlled limit, with $k_2 = 7.7 \times 10^9$ M^{-1} s^{-1}.[3] This radical anion is protonated only in strongly acidic solutions,

$$\cdot COOH = H^+ + \cdot CO_2^- \quad pK_a = 1.4 \tag{2}$$

With a redox potential of -2.0V vs. NHE, the $\cdot CO_2^-$ radical anion is strongly reducing.[4] The decay of the $\cdot CO_2^-$ radical followed second-order kinetics. At pH 7, the second-order rate constant for the decay

11

$$\cdot CO_2^- + \cdot CO_2^- \rightarrow \qquad\qquad (3)$$

was 5×10^8 M^{-1} s^{-1}.[1] The carbonate radical, $\cdot CO_3^-$, was obtained in flash photolysis and pulse radiolysis, by oxidation of carbonate and bicarbonate ions by hydroxyl radicals. These reactions were found to occur with second-order rate constants of 4.2×10^8 and 1.5×10^7 M^{-1} s^{-1}, respectively,[5]

$$\cdot OH + CO_3^{2-} \rightarrow OH^- + \cdot CO_3^- \qquad\qquad (4)$$

$$\cdot OH + HCO_3^- \rightarrow H_2O + \cdot CO_3^- \qquad\qquad (5)$$

The carbonate radical anion was also formed by photoionization of the carbonate ion,[1]

$$CO_3^{2-} + h\nu \rightarrow \cdot CO_3^- + e^- \qquad\qquad (6)$$

Its broad absorption band in the visible region, with $\lambda_{max} = 600$ nm and $\epsilon = 1.9 \times 10^3$ M^{-1} cm^{-1},[6] is convenient for the determination of rate constants of its interactions with other species. The rate constant for the recombination of the $\cdot CO_3^-$ radical has a negligible apparent activation energy, which seems to indicate a composite reaction sequence. The pK_a for the conjugate acid,

$$\cdot CO_3H = \cdot CO_3^- + H^+ \qquad\qquad (7)$$

is between 7.0 and 8.2.[7] The carbonate acts primarily as an oxidant.[1]

In aqueous ethanol solutions containing duroquinone (DQ), the formation of the $\cdot CO_3^-$ radical anion could be determined by the reaction of carbonate ions with the triplet state of duroquinone,

$$DQ_T + \cdot CO_3^{2-} \rightarrow DQ^- + \cdot CO_3^- \qquad\qquad (8)$$

occurring with a rate constant of 7×10^7 M^{-1} s^{-1}, to form durosemiquinone (DQ$^-$).[8] In a micellar solution of dodecyl trimethylammonium chloride in water the above reaction was followed by a subsequent step, in which the $\cdot CO_3^-$ radical reacted with another duroquinone triplet molecule, producing carbon superoxide, which then decomposed to carbon dioxide and oxygen.[9]

$$\cdot CO_3^- + DQ_T \rightarrow CO_3 + DQ^- \qquad\qquad (9)$$

$$CO_3 \rightarrow CO_2 + 1/2O_2 \qquad\qquad (10)$$

The product of the oxidation of CO_2 by the superoxide ion, O_2^-, was shown to be the peroxydicarbonate anion, $C_2O_6^{2-}$. The overall chemical reaction was

$$2O_2^- + 2CO_2 \rightarrow C2O_6^{2-} + O_2 \qquad\qquad (11)$$

for which the electrochemical equivalent was

$$O_2 + 2e^- + 2CO_2 \rightarrow C_2O_6^{2-} \tag{12}$$

The ion was isolated as its tetramethylammonium salt from acetonitrile solution. The rate of the above charge transfer reaction was measured by rotating ring-disc voltammetry in Me_2SO and DMF solutions containing O.1 M tetraethylammonium perchlorate, in the presence of an excess of CO_2 (hence providing pseudo-first-order kinetics). The second-order rate constant thus derived was $k_2 = 1.4 \times 10^3$ M^{-1} s^{-1}. The proposed primary reaction was the nucleophilic addition of the superoxide ion to carbon dioxide, followed by addition of another CO_2 molecule and further electron transfer from another superoxide ion,[10]

$$O_2^- + CO_2 \leftrightarrow {}^-O{-}C(O){-}OO\cdot \tag{13}$$

$$^-O{-}C(O){-}OO\cdot + CO_2 \leftrightarrow {}^-O{-}C(O){-}O{-}C(O){-}OO\cdot \tag{14}$$

$$^-O{-}C(O){-}O{-}C(O){-}OO\cdot + O_2^- \rightarrow {}^-O{-}C(O){-}O{-}C(O){-}OO^- \tag{15}$$

References

1. **Neta, P., Huie, R. E., and Ross, A. B.**, Rate constants for reactions of inorganic radicals in aqueous solution, *J. Phys. Chem. Ref. Data*, 17, 1027–1284, 1988.

2. **Neta, P., Simic, M., and Hayon, E.**, Pulse radiolysis of aliphatic acids in aqueous solutions. I. Simple monocarboxylic acids, *J. Phys. Chem.*, 73, 4207–4213, 1969.

3. **Hart, E. J. and Anbar, M.**, The Hydrated Electron, Wiley-Interscience, New York, 1970.

4. **Butler, J. and Henglein, A.**, Elementary reactions of the reduction of thallium(1 +) in aqueous solution, *Radiat. Phys. Chem.*, 15, 603–612, 1980.

5. **Weeks, J. L. and Rabani, J.**, The pulse radiolysis of deaerated aqueous carbonate solutions. I. Transient optical spectrum and mechanism. II. pK for OH radicals, *J. Phys. Chem.*, 70, 2100–2106, 1966.

6. **Behar, D., Czapski, G., and Duchovny, Y.**, Carbonate radical in flash photolysis and pulse radiolysis of aqueous carbonate solutions, *J. Phys. Chem.*, 74, 2206–2210, 1970.

7. **Eriksen, T. E., Lind, J., and Merenyi, G.**, On the acid-base equilibrium of the carbonate radical, *Radiat. Phys. Chem.*, 26, 197–199, 1985.

8. **Scheerer, R. and Grätzel, M.**, Photoinduced oxidation of carbonate ions by duroquinone, a pathway of oxygen evolution from water by visible light, *Ber. Bunsenges. Phys. Chem.*, 80, 979–982, 1976.

9. **Scheerer, R. and Grätzel, M.**, Laser photolysis studies of duroquinone triplet state electron transfer reactions, *J. Am. Chem. Soc.*, 99, 865–871, 1977.

10. **Roberts, J. L., Jr., Calderwood, T. S., and Sawyer, D. T.**, Nucleophilic oxygenation of carbon dioxide by superoxide ion in aprotic media to form the $C_2O_6{}^{2-}$ species, *J. Am. Chem. Soc.*, 106, 4667–4670, 1984.

FOUR

Coupling Reactions of Carbon Dioxide

CARBAMATES AND CARBONATE ESTERS

The activation of carbon dioxide can be accomplished either by the direct supply of energy — photochemical, electrical, or thermal — or by its reaction with a reactant of high free-energy content. The latter category includes hydrogenation with molecular hydrogen and the reaction with hydrogen-rich molecules, such as ammonia and amines. Probably the earliest example of the scientifically initiated artificial synthesis of an organic compound was Woehler's preparation of urea from ammonium cyanate. Urea, now produced on a large scale from ammonia and carbon dioxide via ammonium carbamate,

$$NH_3 + CO_2 \leftrightarrow H_2NCOOH \tag{1}$$

$$H_2HCOOH + NH_3 \leftrightarrow H_2N\text{--}COO^- \; NH_4^+ \tag{2}$$

is an important chemical intermediate, as well as a valuable nitrogen fertilizer.[1] The equilibrium constant for the above reactions to ammonium carbamate is highly temperature dependent. Its values are 2.08×10^5 at 25°C, 1.11×10^{-5} at 190°C, and 1.14×10^{-6} at 215°C. Thus, at high temperatures, the reaction is very strongly shifted to the left. The reaction may however be shifted to the right by dehydration, producing urea, as in its industrial production,[2]

$$H_2N\text{--}COO^- \; NH_4^+ = H_2NCONH_2 + H_2O \tag{3}$$

A very attractive application of carbonate esters is that of the polycarbonate plastics, which have many important uses.[3]

Another early example was the insertion of carbon dioxide into the C-H bond of benzene by the Friedel-Crafts reaction with anhydrous $AlCl_3$, resulting in benzoic acid. The reaction was preferably carried out at pressures of 50 to 60 atm and temperatures of 80 to 150°C. The carboxylic acid was released by hydrolysis,[4]

$$C_6H_5Al_2Cl_5 + CO_2 \rightarrow C_6H_5CO_2Al_2Cl_5 \qquad (4)$$

$$C_6H_5CO_2Al_2Cl_5 + H_2O \rightarrow C_6H_5COOH + Al_2Cl_5OH \qquad (5)$$

A primary step in the use of waste carbon dioxide, e.g., from coal- or oil-fired power plants, will be to recover and concentrate this gas. Amines, such as mono-, di-, and triethanolamine may be applied for this process, which is already now widely used for removal of carbon dioxide from natural gas. Monoethanolamine and diethanolamine react with carbon dioxide to form carbamates. The carbon dioxide is then released by heating the liquid mixture to 400 K.[5]

The nickel(II) complex of N,N-dimethyl-ethylenediamine, Me_2N-CH_2-CH_2NH_2,

$$[Ni(Me_2N-CH_2-CH_2-NH_2)_3](ClO_4)_2$$

in aqueous ethanol solution was found to achieve the spontaneous fixation of carbon dioxide from air, forming the trinuclear nickel(II) complex

$$[Ni_3(Me_2N-CH_2-CH_2-NH_2)_6(CO_3)(H_2O)_4](ClO_4)_4$$

containing three octahedral Ni(II) atoms. From crystal structure data, the carbonate ligand was shown to be joined by a hydrogen bonding network, which stabilized a bidentate carbonate. The carbonated complex had IR absorption peaks at 1626 and 1377 cm^{-1}, corresponding to the ν_3 vibration of the CO_3^{2-} ion. The carbon dioxide could be released by passing nitrogen gas through a solution of the carbonated complex containing Me_2N-CH_2-CH_2-NH_2 and 1 N ClO_4^- for several hours at 60°C, thus recovering the carbonate-free Ni(II) starting complex.[6]

The iridium complex, $Ir(Cl)(CO)(Ph_3P)_2$, effectively catalyzed the synthesis of formamide from carbon dioxide, ammonia, and hydrogen,

$$CO_2 + H_2 + NH_3 \rightarrow HC(O)NH_2 \qquad (6)$$

carried out in homogeneous toluene or methanol solutions. The yield of formamide increased markedly with rising pressure of the gaseous reactants.[7]

Monoalkylammonium N-alkylcarbamates were easily formed by saturating a solution of the amine in tetrahydrofuran with CO_2,

$$2RNH_2 + CO_2 \rightarrow [RNH_3][O_2CNHR] \qquad (7)$$

in which R = benzyl, allyl, *tert*-butyl, and cyclohexyl. These monoalkyl ammonium carbamates could be made to react with alkyl halides such as allyl bromide
in the presence of crown ethers to produce organic carbamates,

$$[RNH_3][O_2CNHR] + R'X \rightarrow RNHC(O)OR' \tag{8}$$

N-alkylcarbamate esters RNHC(O)OR' were also prepared by direct reaction of
primary amines, carbon dioxide, and alkyl halides in the presence of a macrocyclic
polyether, such as 18-crown-6. This method offers an attractive alternative route
to the important group of carbamate esters, not requiring the use of the very toxic
phosgene or alkyl isocyanates.[8]

In methanol solutions, carbon dioxide forms with primary and secondary amines
quite stable carbamic acid amine salts, i.e., the equilibrium in the equation,

$$CO_2 + 2R^1R^2NH = R^1R^2NCOO^- \, (R^1R^2NH_2)^+ \tag{9}$$

is very much to the right. Primary aliphatic amines and carbon dioxide in methanol
solution reacted with oxiranes, such as 2-aryl-2-methoxy 3-methyloxiranes, yielding the corresponding 4-aryl-4-hydroxy 5-methlyloxazolidin-2-one derivatives,[9]

Carbon dioxide was activated at room temperature and atmospheric pressure by
(5,10,15,20-tetraphenyl porphinato) aluminum acetate in the presence of secondary amines and epoxides, producing dialkylcarbamic esters. The Al-porphyrin
acted catalytically. Thus, with 1,2-epoxypropane and diethylamine, CO_2, and the
Al-porphyrin, the turnover number with respect to the formation of 2-hydroxy-
propyl diethylcarbamate was six. Carbamic esters were also obtained from other
dialkylamines, such as piperidine and diisopropylamine. In the proposed mechanism, CO_2 was trapped in the form of an Al-carbamate on the Al-porphyrin, on
the opposite side of the acetate group with respect to the porphyrin plane.[10]

In a route to carbamates which avoids the intermediate production of the very
toxic phosgene and isocyanates, a palladium-complex catalyzed synthesis was
developed, using amines, carbon dioxide, and cyclic diolefins as starting com-

ponents. Carbamates are important as insecticides and as intermediates for specialty chemical applications. The reaction depended on (1) the activation of CO_2 by a primary or secondary amine, carried out in THF or methylene solution, in the presence of a tertiary amine base such as quinuclidine,[11]

$$RR'NH + Base + CO_2 = RR'NCO_2^- \ ^+HBase \qquad (10)$$

(2) treatment of a cyclic diolefin, such as norbornadiene, or dicyclopentadiene, or 1,5-cyclooctadiene with palladium dichloride, thus activating the diolefin by the Pd-metal center toward nucleophilic attack by the carbamate anion. This reaction of the carbamate salt with the Pd-diolefins was performed in a CO_2 atmosphere at low temperatures. (3) The resulting metal-carbon bond was then cleaved by protonolysis with DIPHOS [bis(diphenylphosphino)ethane] followed by $NaBH_4$ to produce the corresponding carbamate esters. Thus, norbornadiene-palladium chloride with various carbamates formed nortricyclo carbamate esters.[11]

Urea derivatives are important as agricultural pesticides, such as in uron herbicides, and also as pharmaceuticals. The common methods of synthesis involve the very toxic phosgene and isocyanates as starting materials and intermediates. A direct route to ureas was achieved using aliphatic and araliphatic primary amines, which reacted with CO_2 to form ammonium carbamates. In the presence of ruthenium complexes and a terminal alkyne,

$$Ru[H\text{–}C \equiv C\text{–}R']$$

an *in situ* reaction occurred, producing N,N'-disubstituted symmetrical ureas,[12]

$$2RNH_2 + CO_2 \rightarrow RNH\text{–}C(O)\text{–}NHR + H_2O \qquad (11)$$

In this reaction, the alkyne ruthenium intermediate acted as a dehydrating agent. Best results were obtained with a catalyst precursor from $RuCl_3 \cdot 3H_2O$ and *n*-Bu_3P, with 2-methylbut-3-yn-2-ol as the alkyne. The mixture was pressurized with CO_2 to 5 mPa, and heated to 140°C. With cyclohexylamine as the primary amine, N,N'-dicyclohexylurea was produced during 20 h in up to 62% yield.[13] With secondary amines and terminal alkynes, the course of the reaction was different, producing vinyl carbamates,[13]

$$CH \equiv CR' + R2NH + CO2 \xrightarrow{[Ru]}$$

$$R2NCOO{-}CH = CHR'$$

Secondary amines and CO_2 reacted with propargyl alcohols, in the presence of ruthenium complexes, producing β-oxopropyl carbamates,[14]

α-Aminonitriles reacted with CO_2, neat, at room temperature, yielding disubstituted ureas. Subsequent reactions with water at room temperature yielded N-(3)-substituted hydantoins in yields of 80 to 100%,[15]

The reaction of carbon dioxide with tertiary amines, such as triethanolamine (TEA), is widely used in industry, particularly for natural-gas purification. In contrast to primary and secondary amines, tertiary amines do not form carbamates. Natural-gas washing for removal of carbon dioxide with aqueous solutions of triethanolamine has the added advantage of also removing hydrogen sulfide from the gaseous phase. In a kinetic study of the reaction of carbon dioxide with two tertiary amines — TEA and methyldiethanolamine (MDEA) — in aqueous solutions, the stopped flow method was used, with a color indicator to reveal the pH change due to the formation of the bicarbonate ion. Essentially, the amines catalyzed the hydration of carbon dioxide.[16]

Carbonate esters may in principle be obtained by the reaction of carbon dioxide with alcohols, but the equilibrium is in favor of the dissociation of the esters,

$$ROH + CO_2 = ROCOOH \qquad (12)$$

The carbonate esters can however be fixed by reaction with epoxides, producing hydroxylalkyl alkyl carbonates. With cyclohexene oxide and methanol, in an autoclave charged with carbon dioxide to 50 atm pressure, at a temperature of 120°C for 21 h, followed by an acetylation step, the yield of 2-acetoxy cyclohexyl methyl carbonate (a) was about 3%. With increased reaction temperature, to 160°C, the yield of 2-methoxycyclohexyl acetate (b) reached 70%,[17]

CO$_2$ — METAL CENTER COORDINATION

Carbon dioxide, while by itself rather inert, may be activated by some reagent, usually a transition element compound, which enables coupling of carbon dioxide with suitable organic molecules. Thus, condensation products are formed containing the CO_2 moiety.[18] Various compounds can be produced by nonreductive fixation of carbon dioxide, such as its condensation with ethylene oxide to produce ethylene carbonate and with sodium phenolate to yield acetylsalicylic acid.

Reactions of carbon dioxide coupling include the coordination of CO_2 to transition metal complexes, with insertion into metal-carbon bonds, oxidative coupling with CO_2, and insertion reactions of CO_2. In such catalytic reactions, carbon dioxide undergoes condensation with alkynes, alkenes, dienes, and benzene derivatives, forming pyrones, lactones, esters, and carboxylic acids — using complexes of ruthenium, rhodium, nickel, or palladium for activation of carbon dioxide.[19-27]

One case in which the carbon dioxide-metal complex was structurally characterized was that of the bis(tricyclophosphine) nickel complex, $[Ni(CO_2)(PCy_3)_2, 0.75(C_7H_8)]$, in which Cy = cyclohexyl. The complex was prepared by reacting $[Ni(PCy_3)_3]$ with carbon dioxide in toluene. The complex was shown by IR and single-crystal study to be planar, with the CO_2 ligand in bent geometry, coordinated to the Ni atom by the carbon atom and one of the oxygen atoms, thus forming the η^2-C,O mode of bonding CO_2 to a metal center,[28,29]

$$M \overset{O}{\underset{C}{\diagdown}} \overset{\diagup}{\diagdown} _O \qquad M = Ni, Nb, Ta, Mo, V$$

A different type of bonding, presumably by η^1-C bonding of CO_2 to an axial metal coordination site,

$$M - C \overset{\diagup O}{\underset{O}{\diagdown}}$$

was found in the CO_2 complexes with metal macrocycles, such as in the 1:1 complexes with cobalt macrocycles $[Co(I)L_5]^+$ (L_5 in these complexes is 5,7,7,12,14,14-hexamethyl 1,4,8,11 tetraaza cyclotetradeca-4,11-diene),

With the racemic form of this cobalt macrocycle, in acetonitrile or dimethyl sulfoxide solutions, the CO_2 binding constant was $6 \times 10^2 \, M^{-1}$. The $[CoL_5CO_2]^+$ complex in acetonitrile solution underwent second-order decomposition, yielding CO and HCO_3^- as stable products.[30,31]

Another example was the isolation of a crystalline carbonato complex of bis (cyclo-pentadienyl) hydridophenyl tungsten(IV), $Cp_2W(H)Ph$ (where $Cp = \eta^5$-C_5H_5), which reacted with CO_2 in wet acetone to form the mononuclear carbonato complex $Cp_2W(\eta^2$-$CO_3)$. The product was characterized by IR, ^1H-, and ^{13}C-NMR spectroscopy, and by X-ray diffraction.[32]

$$Cp_2W(H)Ph + CO_2 + H_2O + Me_2CO \rightarrow Cp_2W(\eta^2\text{--}CO_3) + Me_2CHOH \qquad (13)$$

Carbon dioxide reacted with low-valent cobalt and rhodium complexes $(np_3)CoH$ and $(np_3)RhH$, where $np_3 =$ tris(2-diphenylphosphino)ethylamine, in the presence of a solvated or complexed Lewis acid, such as the Na^+ ion. On adding a tetrahydrofuran solution of $NaBPh_4$ to a solution of $(np_3)CoH$ under a CO_2 atmosphere, the solution turned red-brown, with formation of the carbonyl complex $[(np_3)Co(CO)]BPh_4$ in 50% yield. Similarly, $(np_3)RhH$ reacted with CO_2 in the presence of Na^+ ions to produce the diamagnetic complex $[(np_3)Rh(CO)]BPh_4$ in 50% yield.[33]

The cobalt complex tris(2,2'-bipyridine)-cobalt(I), $Co(bpy)_3^+$, in aqueous bicarbonate solutions (pH 8.5 to 10) reacted to form carbon monoxide. The CO produced was scavenged by additional $Co(bpy)_3^+$, precipitating the insoluble $[Co(bpy)(CO)_2]_2$. A proposed intermediate in the reaction is $Co(bpy)_2(H_2O)H^{2+}$.[34]

CO_2 COUPLING TO ORGANIC COMPOUNDS

The long-known reaction of carbon dioxide with Grignard reagents is a useful tool for regiospecific carboxylation, a which the driving force is the oxidation of one magnesium atom for each alkyl halide converted,[29]

$$R-X + Mg \rightarrow RMgX \qquad (14)$$

$$RMgX + CO_2 \rightarrow R-COOMgX \qquad (15)$$

$$R-COOMgX + H_2O \rightarrow R-COOH + MgX(OH) \qquad (16)$$

Pioneering work in the field of the coupling of CO_2 into organic molecules was that of reacting butadiene with carbon dioxide, in the presence of Pd(O)-phosphine complexes, such as $Pd(Ph_2PCH_2CH_2PPh_2)_2$, in polar aprotic solvents such as DMF, with CO_2 at 50 atm pressure, at 120°C. The products were mainly butadiene oligomers (about 60%), as well as lactones (12%).[35,36] In the telomerization of butadiene with CO_2, palladium complexes were found to act as specific CO_2 carriers, such as in the reaction

The main product of the telomerization reaction with butadiene was the six-membered cyclic lactone,[26]

as well as carboxylic acids and esters. This telomerization reaction of butadiene produced very valuable compounds, useful as monomers or intermediates in the manufacture of polyester resins, pesticides, and plasticizers.[18,24]

In the carbon dioxide molecule, the oxygen atoms act as weak Lewis bases, while the carbon atom is a weak Lewis acid. Thus, CO_2 can react in the coordination sphere of transition metal complexes, e.g., by inserting into metal-oxygen, metal-carbon, metal-hydrogen, or metal-nitrogen bonds, forming products with one or more carbon atoms. One group of such transition metal complexes which has been studied carefully is that of the cobalt macrocycle CoL^+ (L = [14]diene-5,7,7,12,14,14-hexamethyl-1,4,8,11-tetraazacyclotetradeca-4,1-diene),

CoL^{2+} in acetonitrile-H_2O solutions was reported to catalyze the electrochemical reduction of carbon dioxide to carbon monoxide[37] (see also Chapter 7). In dry DMSO[38] and in acetonitrile[39] the binding of CoL^+ with CO_2 was shown to be reversible, according to the equation,

$$Co([14]diene)^+ + CO_2 = Co([14]diene)(CO_2)^+ \qquad \textbf{(17)}$$

Regioselective synthesis of O-1-(1,3-dienyl) carbamates was obtained by the addition of secondary amines and CO_2 to isopropenylacetylene in the present of a ruthenium-containing catalyst precursor, containing a chelating bidentate phosphine ligand, such as $[Ph_2P(CH_2)_nPPh_2]Ru(\eta^3\text{-}CH_2 = C(Me)CH_2)_2$, where n = 1 to 4,

When n = 2, with the secondary amines morpholine, pyrrolidine, piperidine, and diethylamine, the yields of the O-1-(1,3-dienyl)carbamates were 62, 50, 36, and 31%, respectively. Such dienyl carbamates are very useful as substrates for the Diels-Alder reaction and as precursors for polymers and fiber.[40]

Nickel(O)-catalyzed fixation of carbon dioxide into diynes such as 1,6-heptadiyne or 1,7-octadiyne caused intramolecular cyclization of the diynes, producing bicyclic 2-pyrones fused with five- and six-membered carbocycles.[41]

Alternating copolymerization, in which carbon dioxide serves as a comonomer, has been achieved by intermolecular cyclization of acyclic or cyclic diynes, in the presence of nickel(O) catalysts, producing poly-(2-pyrones). 3,11-Tetradecadiyne was copolymerized with carbon dioxide under pressure in a solvent mixture of THF-MeCN at 110°C in the presence of the Ni(O) catalyst obtained from $Ni(COD)_2$ and a tertiary phosphine, such as PEt_3.[42]

A catalyst generated from a mixture of Pd_2(dibenzylidene acetone).$CHCl_3$ and 1-(2-pyridyl)-2-(di-n-butyl-phosphino)ethane effected the cyclo-addition of methoxyallene with CO_2 in acetonitrile solution under pressure at 120°C to produce (E)-5-methoxy-2-(methoxy-methylene)-4-methylene-5-pentanolide in a regio- and stereospecific reaction, in yields of up to 64%,[39]

An interesting approach to the incorporation of carbon dioxide into organic molecules is that of plasma polymerization, in which thin films are formed by activation of monomer molecules in a radiofrequency induced plasma. Mixtures of carbon dioxide and perfluorobenzene or benzene were excited in an inductively coupled system operated at 13.56 MHz. With the perfluorobenzene-CO_2 mixture, the plasma reaction produced films with carboxylic acid groups, which may have useful ion-exchange properties. With the benzene-CO_2 mixture, the resulting film contained oxygen atoms, but no carboxylic acid groups.[44]

The reaction of derivatives of propargylic alcohol with carbon dioxide in the presence of various transition metal complexes led to α-methylene cyclic carbonates. Thus, cycloaddition of CO_2 with 2-methyl-3-butyn-2-ol resulted in the cyclic carbonate,[45]

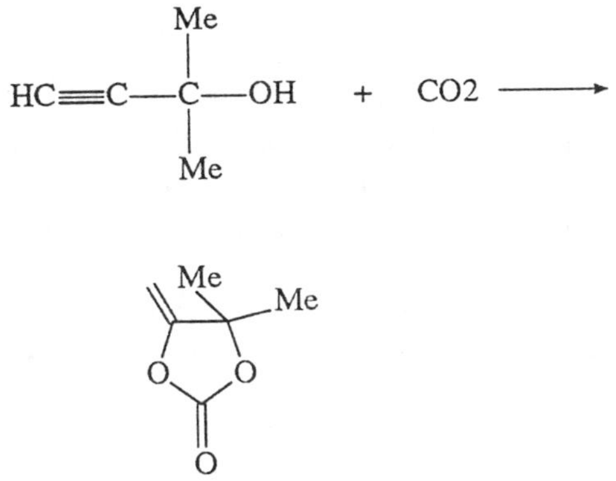

$$\text{HC} \equiv \text{C} - \underset{\underset{\text{Me}}{|}}{\overset{\overset{\text{Me}}{|}}{\text{C}}} - \text{OH} \quad + \quad \text{CO2} \longrightarrow$$

In the presence of iodobenzene and $Pd(PPh_3)_4$ as catalyst, sodium 2-methyl-3-butyn-2-olate reacted with pressurized CO_2 (10 atm) at 100°C to produce stereoselectivity in 68% yield the cyclic vinylidene carbonate (E)-5-benzylidene-4,4-dimethyl-1,3-dioxolan-2-one,[46]

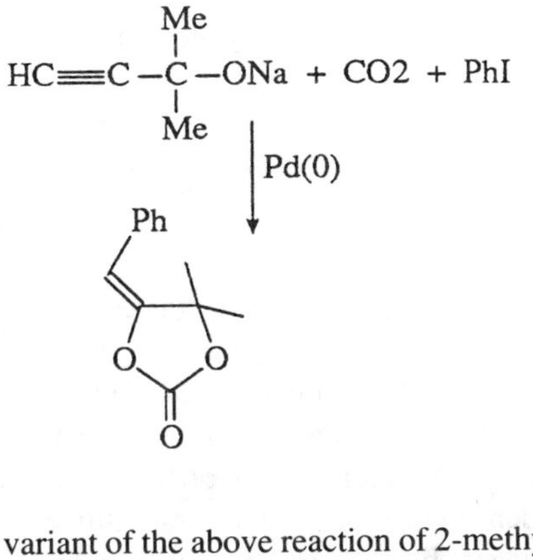

$$\text{HC} \equiv \text{C} - \underset{\underset{\text{Me}}{|}}{\overset{\overset{\text{Me}}{|}}{\text{C}}} - \text{ONa} + \text{CO2} + \text{PhI}$$

Pd(0)

A most interesting variant of the above reaction of 2-methyl-3-butyn-2-ol occurred in the presence of carbon monoxide, carbon dioxide, iodobenzene, triethylamine, and transition metal-PPh_3 complexes as catalysts, such as $Pt(PPh_3)_4$. In this reaction, carbon monoxide entered into the 2-methyl-3-butyn-2-ol molecule to produce 2,2-dimethyl-5-phenyl-3(2H)-furanone,[45,47]

$$\text{HC}\equiv\text{C}-\underset{\underset{\text{Me}}{|}}{\overset{\overset{\text{Me}}{|}}{\text{C}}}-\text{OH} + \text{CO} + \text{Ph}-\text{I} \xrightarrow{\text{CO}_2}$$

This compound is known as bullatenone, occurring in the essential oil of *Myrtus bullata*, a shrub which is endemic in New Zealand. The 3(2H)-furanone ring system is a central component of several antitumor agents and also a constituent of some flavoring compounds.[45,47]

The 1:1 addition compound of triphenylphosphine with dimethyl acetylenedicarboxylate reacted with carbon dioxide to form a mixture of two carboxylate betains,[48]

With trialkyl phosphite instead of triphenyl phosphine, a different reaction took place with dimethyl acetylenedicarboxylate and carbon dioxide, producing a non-ionic cyclic phosphorane,[49]

Reversible fixation of carbon dioxide was achieved by bifunctional complexes containing in their structure a nucleophilic cobalt(I) and an alkali cation. Complexes exhibiting such fixation were Co(R-salen)M (where salen = N,N′-ethylene-bis(salicylideneaminato), R-salen = substituted salen ligand, and M = Li, Na, K, Cs). In solvents like tetrahydrofuran, pyridine, or toluene, the deep-green solutions of the bimetallic complexes [Co(R-salen)M] reversibly absorbed 1 mol of CO_2 per bimetallic unit. X-ray analysis of the adduct [Co(R-salen)MCO$_2$] indicated that CO_2 was anchored to the cobalt atom through a Co-C σ bond, while the oxygen atoms interacted with the alkali cation[50] (see also Chapter 7 for the structure of Co-salen and for its use as mediator in the electrochemical reduction of CO_2).

Carbon dioxide reacted with hydrido-chloro-bis-(cyclopentadienyl)-zirconium(IV), $cp_2Zr(H)(Cl)$, by insertion into the Z-H bond, followed by release of formaldehyde,

$$cp_2Zr(H)(Cl) + CO_2 \rightarrow cp_2(Cl)Zr-O-Zr(Cl)cp_2 + HCHO \qquad (18)$$

In the presence of an excess of the starting complex, the formaldehyde was reduced and trapped as a methoxy ligand.[51]

$$cp_2Zr(H)(Cl) + HCHO \rightarrow cp_2Zr(OCH_3)(Cl) \qquad (19)$$

Carbon dioxide at normal pressure, in the presence of 2,2'-dipyridyl-cyclooctadiene-1,5-nickel(O), (dipy)Ni(COD), in the THF solution at $-10°C$, reacted with bis-cyclohexyl-carbodiimide, $Cy-N=C=N-Cy$, to form a five-membered nickel chelate,[52]

$$dipyNi(COD) + CO_2 + Cy-N=C=N-Cy \longrightarrow COD +$$

Reduction of carbon dioxide by incorporation into a bimetallic polyhydride complex was carried out using the rhodium-osmium complex $(COD)RhH_3$-OsP_3, where $COD = $ 1,5-cyclooctadiene and $P_3 = (PMe_2Ph)_3$. This complex in the THF solution reacted completely with an excess of CO_2 (1 atm) within 8 h at 25°C, producing two complexes, as described by the stoichiometry,[53]

$$2(COD)RhH_3OsP_3 + 2CO_2 \rightarrow (COD)_2Rh_2OsP_3H_2CO_2$$
$$+ H_2Os(CO)P_3 + H_2O \qquad (20)$$

In the one product, $H_2Os(CO)P_3$, CO_2 had been reduced to the CO moiety. The other product, $(COD)_2Rh_2OsP_3H_2CO_2$, is a unique example of a neutral compound containing hydride as well as CO_2 ligands.[53]

Polypropylene carbonate (PPC), which is obtained by the copolymerization of propylene oxide with carbon dioxide, could be a very useful polymeric material.

Its drawback hitherto has been insufficient thermal stability. Thus, as tested by thermogravimetry, already at 215°C, there occurred a 5% weight loss.[54] This thermal degradation was found to be due to an "unzipping" of the polymer, releasing propylene carbonate. In an effort to overcome this thermal instability, the effects of various added monomers were studied. Considerably enhanced thermal stability was attained by using maleic anhydride as an added monomer. Using propylene oxide (PO) and maleic anhydride (MA), at an MA/PO ratio of 0.72, the temperature of 5% weight loss was raised to 263°C. The catalyst used for these polymerizations was a polymer-chelated bimetallic cyanide, (polymer) $[Zn(Fe(CN)_6]_xCl_{2-3x}$.[55]

References

1. **Mavrovic, I. and Shirley, A. R.,** Urea, in *Kirk-Othmer Encyclopedia of Chemical Technology,* Vol. 23, 3rd ed., Wiley-Interscience, New York, 1983, 548–575.

2. **Klier, K.,** Catalytic conversions of CO_2: activation and hydrogenation, Proc. Int. Symp. Chemical Fixation of Carbon Dioxide, Nagoya, Japan, Dec. 2–4, 1991, 139–134.

3. **Fox, D. W.,** Polycarbonates, in *Kirk-Othmer Encyclopedia of Chemical Technology,* Vol. 18, 3rd ed., Wiley-Interscience, New York, 1982, 479–494.

4. **Thomas, C. A.,** *Anhydrous Aluminum Chloride in Organic Chemistry,* Reinhold, New York, 1941, 508.

5. **Ballou, W. R.,** Carbon dioxide, in *Kirk-Othmer Encyclopedia of Chemical Technology,* Vol. 4, 3rd ed., Wiley-Interscience, New York, 1978, 725–742.

6. **Tanase, T., Nitta, S., Yoshikawa, S., Kobayashi, K., Sakurai, T., and Yano, S.,** Spontaneous fixation of carbon dioxide in air by a nickel diamine complex: synthesis and characterization of a trinuclear nickel(II) complex with a novel hydrogen bonding system around a carbonate ligand, *Inorg. Chem.,* 31, 1058–1062, 1992.

7. **Vaska, L., Schreiner, S., Felty, R. A., and Yu, J. Y.,** Catalytic reduction of carbon dioxide to methane and other species via formamide intermediation: synthesis and hydrogenation of $HC(O)NH_2$ in the presence of $[Ir(Cl)(CO)(Ph_3P)_2]$, *J. Mol. Catal.,* 52, 11–16, 1989.

8. **Aresta, M. and Quaranta, E.,** Role of the macrocyclic polyether in the synthesis of N-alkylcarbamate esters from primary amines, CO_2, and alkyl halides in the presence of crown-ethers, *Tetrahedron,* 48, 1515–1530, 1992.

9. **Toda, T., Yoshida, M., Ohshima, M., Yagi, K., and Komatsu, S.,** Carbon dioxide fixation reactions via carbamic acid amine salts, Proc. Int. Symp. Chemical Fixation of Carbon Dioxide, Nagoya, Japan, Dec. 2-4, 1991, 185–188.

10. **Kojima, F., Aida, T., and Inoue, S.,** Fixation and activation of carbon dioxide on aluminum porphyrin. Catalytic formation of carbamic ester from carbon dioxide, amine and epoxide, *J. Am. Chem. Soc.,* 108, 391–395, 1986.

11. **McGhee, W. D. and Riley, D. P.,** Palladium-mediated synthesis of urethanes from amines, carbon dioxide, and cyclic diolefins, *Organometallics,* 11, 900–907, 1992.

12. **Fournier, J., Bruneau, C., Dixneuf, P. H., and Lécolier, S.,** Ruthenium-catalyzed synthesis of symmetrical N,N'-dialkylureas directly from carbon dioxide and amines, *J. Org. Chem.,* 56, 4456–4458, 1991.

13. **Mahé, R., Sasaki, Y., Bruneau, C., and Dixneuf, P. H.,** Catalytic synthesis of vinyl carbamates from carbon dioxide and alkynes with ruthenium complexes, *J. Org. Chem.,* 54, 1518–1523, 1989.

14. **Sasaki, Y. and Dixneuf, P. H.**, Ruthenium-catalyzed reaction of carbon dioxide, amine and acetylenic alcohol, *J. Org. Chem.*, 52, 4389–4391, 1987.

15. **O'Brien, R. A., Worman, J. J., and Olson, E. S.**, Carbon dioxide in organic synthesis. Preparation and mechanism of formation of N-(3)-substituted hydantoins, *Synth. Commun.*, 22, 823–828, 1992.

16. **Barth, D., Tondre, C., Lappai, G., and Delpuech, J.-J.**, Kinetic study of carbon dioxide reaction with tertiary amines in aqueous solutions, *J. Phys. Chem.*, 85, 3660–3667, 1981.

17. **Yoshida, Y. and Ishii, S.**, A novel synthesis of carbonate ester from the reaction of CO_2, alcohol and epoxide, Proc. Int. Symp. Chemical Fixation of Carbon Dioxide, Nagoya, Japan, Dec. 2–4, 1991, 423–426.

18. **Braunstein, P., Matt, D., and Nobel, D.**, Reactions of carbon dioxide with carbon-carbon bond formation catalyzed by transition metal complexes, *Chem. Rev.*, 88, 747–764, 1988.

19. **Inoue, S. and Yamazaki, N.**, *Organic and Bio-Organic Chemistry of Carbon Dioxide*, Kodansha, Tokyo, 1982.

20. **Ziessel, R.**, Chimie de coordination de la molecule de dioxyde de carbon: activation biologique, chimique, electrochimique et photochimique, *Nouv. J. Chim.*, 7, 613–633, 1983.

21. **Behr, A.**, The synthesis of organic chemicals by catalytic reactions of carbon dioxide, *Bull. Soc. Chim. Belg.*, 94, 671–683, 1985.

22. **Behr, A.**, Use of carbon dioxide in industrial organic synthesis, *Chem. Eng. Technol.*, 10, 16–27, 1987.

23. **Behr, A.**, Carbon dioxide as an alternative C_1 synthetic unit: activation by transition-metal complexes, *Angew. Chem. Int. Ed. Engl.*, 27, 661–678, 1988.

24. **Behr, A.**, Carbon dioxide as building block for fine chemicals synthesis, in *Aspects of Homogeneous Catalysis*, Ugo, R., Ed., D. Reidel, Dordrecht, The Netherlands, 1988, 59–96.

25. **Aresta, M. and Forti, G., Eds.**, *Carbon Dioxide as a Source of Carbon. Biochemical and Chemical Use*, Kluwer Academic, Hague, 1987.

26. **Braunstein, P., Matt, D., and Nobel, D.**, Carbon dioxide activation of catalytic lactone synthesis by telomerization of butadiene and CO_2, *J. Am. Chem. Soc.*, 110, 3207–3212, 1988.

27. **Kubiak, C. P. and Ratliff, K. S.**, Approaches to the chemical, electrochemical and photochemical activation of carbon dioxide by transition metal complexes, *Israel J. Chem.*, 31, 3–15, 1991.

28. **Aresta, M., Nobile, C. F., Albano, V. G., Forni, E., and Manassero, M.**, New nickel-carbon dioxide complex: synthesis, properties and crystallographic characterization of (carbon dioxide)-bis(tricyclophosphine) nickel, *J. Chem. Soc. Chem. Commun.*, 1975, 636–637.

29. **Aresta, M., Quaranta, E., and Tommasi, I.**, The role of metal centres in reduction and carboxylation reactions utilizing carbon dioxide, Proc. Int. Symp. Chemical Fixation of Carbon Dioxide, Nagoya, Japan, Dec. 2–4, 1991, 209–226.

30. **Fujita, E., Creutz, C., Sutin, N., and Szalda, D. J.**, Carbon dioxide activation by cobalt macrocycles — factors affecting CO_2 and CO binding, *J. Am. Chem. Soc.*, 113, 343–353, 1991.

31. **Fujita, E., Creutz, C., and Sutin, N.**, Carbon dioxide activation by metal macrocycles, Proc. Int. Symp. Chemical Fixation of Carbon Dioxide, Nagoya, Japan, Dec. 2–4, 1991, 243–246.

32. **Ito, T., Sugimoto, S., Ohki, T., Nakano, T., and Osakada, K.**, Reaction of bis (cyclopentadienyl) hydridophenyl tungsten(IV) with carboxylic acids, carbon dioxide, and related compounds, *J. Organometal. Chem.*, 428, 69–83, 1992.

33. **Bianchini, C. and Meli, A.**, Bifunctional activation of CO_2: a case where the acidic and basic sites are not held in the same structure, *J. Am. Chem. Soc.*, 106, 2698–2699, 1984.

34. **Keene, F. R., Creutz, C., and Sutin, N.**, Reduction of carbon dioxide by tris (2,2'-bipyridine) cobalt(I), *Coord. Chem. Rev.*, 64, 247–260, 1985.

35. **Sasaki, Y., Inoue, Y., and Hashimoto, H.**, Reaction of carbon dioxide with butadiene catalysed by palladium complexes. Synthesis of 2-ethylidene-5-en-4-olide, *J. Chem. Soc.*, 1976, 605–606.

36. **Inoue, Y., Sasaki, Y., and Hashimoto, H.**, Incorporation of CO_2 in butadiene dimerization catalyzed by palladium complexes, *Bull. Chem. Soc. Jpn.*, 51, 2375–2378, 1978.

37. **Fisher, B. J. and Eisenberg, R.**, Electrocatalytic reduction of carbon dioxide by using macrocycles of nickel and cobalt, *J. Am. Chem. Soc.*, 102, 7361–7363, 1980.

38. **Gangi, D. A. and Durand, R. R.**, Binding of carbon dioxide to cobalt and nickel tetraaza-macrocycles, *J. Chem. Soc. Chem. Commun.*, 1986, 697–699.

39. **Fujita, E., Szalda, D. J., Creutz, C., and Sutin, N.**, Carbon dioxide activation: thermodynamics of CO_2 binding and the involvement of the two cobalt centers in the reduction of CO_2 by a cobalt(I) macrocycle, *J. Am. Chem. Soc.*, 110, 4870–4871, 1988.

40. **Höfer, J., Doucet, H., Bruneau, C., and Dixneuf, P. H.**, Ruthenium catalysed regioselective synthesis of O-1(1,3-dienyl) carbamates directly from CO_2, *Tetrahedr. Lett.*, 32, 7409–7410, 1991.

41. **Tsuda, T., Morikawa, S., Hasegawa, H., and Saegusa, T.**, Nickel(O)-catalyzed cycloaddition of silyl diynes with carbon dioxide to silyl bicyclic α-pyrones, *J. Org. Chem.*, 55, 2978–2981, 1990.

42. **Tsuda, T., Maruta, K., and Kitaike, Y.**, Nickel (O)-catalyzed alternating copolymerization of carbon dioxide with diynes to poly(2-pyrones), *J. Am. Chem. Soc.*, 114, 1498–1499, 1992.

43. **Tsuda, T., Yamamoto, T., and Saegusa, T.**, Palladium-catalyzed cycloaddition of carbon dioxide with methoxyallene, *J. Organometal. Chem.*, 429, 46–48, 1992.

44. **Inagaki, N., Tasaka, S., and Chengfei, Z.**, Preparation of thin films containing sulfonic acid and carboxylic acid groups by plasma polymerization of perfluorobenzene/sulfur dioxide and perfluorobenzene/carbon dioxide mixtures, *Polymer Bull.*, 26, 187–191, 1991.

45. **Inoue, Y.**, Synthesis of heterocyclic compounds from CO_2 and propargylic alcohol catalyzed by transition metal complexes, Proc. Int. Symp. Chemical Fixation of Carbon Dioxide, Nagoya, Japan, Dec. 2–4, 1991, 273–280.

46. **Inoue, Y., Itoh, Y., Yen, I.-F., and Imaizumi, S.**, Palladium (O) catalyzed carboxylative cyclyzed coupling of propargylic alcohol with aryl halides, *J. Mol. Catal.*, 60, 1–3, 1990.

47. **Inoue, Y., Ohuchi, K., Yen, I.-F., and Imaizumi, S.**, Preparation of 3(2H)-furanones from 2-propynyl alcohol, CO, and phenyl halides under CO_2 atmosphere catalyzed by transition metal complexes, *Bull. Chem. Soc. Jpn.*, 62, 3518–3522, 1989.

48. **Johnson, A. W. and Tebby, J. C.**, The adducts from triphenylphosphine and dimethyl acetylene dicarboxylate, *J. Chem. Soc.*, 1961, 2126–2130.

49. **Griffiths, D. and Tebby, J. C.**, Fixation and deoxygenation of carbon dioxide to form furans using organophosphorus intermediates, *J. Chem. Soc. Chem. Commun.*, 1981, 607–608.

50. **Gambarotta, S., Arena, F., Floriani, C., and Zanazzi, P. F.,** Carbon dioxide fixation: bifunctional complexes containing acidic and basic sites working as reversible carriers, *J. Am. Chem. Soc.,* 104, 5082–5092, 1982.

51. **Gambarotta, S., Strologo, S., Floriani, C., Chiesi-Villa, A., and Guastini, C.,** Stepwise reduction of CO_2 to formaldehyde and methanol: reaction of CO_2 and CO_2-like molecules with hydrido chloro bis (cyclopentadienyl) zirconium (IV), *J. Am. Chem. Soc.,* 107, 6278–6282, 1985.

52. **Walther, D., Dinjus, E., and Herzog, V.,** Activation of carbon dioxide at transition metal centers — the oxidative coupling of CO_2 with carbodiimides at electron rich nickel(O), *Zeitschr. Chem.,* 23, 188, 1983.

53. **Lundquist, E. G., Huffman, J. C., and Caulton, K. G.,** Formation of a heterometallic carbon dioxide complex with concurrent reduction of CO_2, *J. Am. Chem. Soc.,* 108, 8309–8310, 1986.

54. **Dixon, D. D., Ford, M. E., and Mantell, G. J.,** Thermal stabilization of poly(alkylenecarbonates), *J. Polym. Sci. Polym. Lett.,* 18, 131–134, 1980.

55. **Chen, L.-B., Yang, S.-Y., and Peng, H.,** Improving thermostability of CO_2-epoxide copolymers, *Proc. Int. Symp. Chemical Fixation of Carbon Dioxide,* Nagoya, Japan, Dec. 2–4, 1991, 253–258.

FIVE

Thermal Heterogeneous Reduction

The hydrogenation of carbon dioxide may lead to several reactions, depending on the catalysts used: (1) the reverse water gas shift reaction, producing carbon monoxide and water,

$$CO_2(g) + 4H_2(g) = CO(g) + H_2O(g) \qquad (1)$$

(2) Methanation, which is the reverse of the steam reforming reaction,

$$CO_2(g) + 4H_2(g) = CH_4(g) + 2H_2O(g) \qquad (2)$$

and (3) methanol production,

$$CO_2 + 3H_2 = CH_3OH + H_2O \qquad (3)$$

Other important catalytic reactions of carbon dioxide are (4) the *reforming* of methane and other hydrocarbons to produce synthesis gas,

$$CH_4 + CO_2 = 2CO + 2H_2 \qquad (4)$$

and (5) the gasification of carbon by carbon dioxide,

$$CO_2 + C = 2CO \qquad (5)$$

CARBON DIOXIDE REFORMING

For the conversion of hydrocarbon feedstocks such as natural gas (mainly methane), petroleum, and coal to chemical intermediates, an important step is that of *re-forming*. Reforming refers to the highly endothermic catalytic conversion of such hydrocarbons with either carbon dioxide or steam to produce synthesis gas, a mixture of carbon monoxide, carbon dioxide, and hydrogen. Synthesis gas may be converted by various exothermic reactions to very useful products, such as methanol, ethanol, ethylene, and acetone.

The interaction of carbon dioxide with the widely available methane to form synthesis gas,

$$CH_4 + CO_2 = 2CO + 2H_2 \tag{6}$$

is most attractive for converting the two inexpensive raw materials into more valuable products. A commercial application of this reaction has been developed into the "Calcor Process", a multistage process of reacting CO_2 with natural gas, liquid petroleum gas (LPG), and syngas, to produce high-purity carbon monoxide.[1,2] In a study of this reaction (with CH_4/CO_2 ratio 1:1, at space velocities of 3000 to 6000 h^{-1}) over alumina-supported platinum metals, in the temperature range 723 to 823 K, the specific activities of the catalysts were found to decrease in the order Ru, Pd, Rh, Pt, and Ir. This order of activity was explained to be related to the order for the dissociation of carbon dioxide to oxygen,

$$CO_2 \rightarrow CO + O \tag{7}$$

Presumably the surface-adsorbed O atoms thus formed promote the dissociation of surface-adsorbed methane. The reaction products were mainly carbon monoxide, hydrogen, water, carbon deposits on the catalysts, and traces of ethane. The ratio of the products indicates the participation of secondary reactions, such as the methanation of carbon dioxide and the reverse water gas shift reaction.[2]

Very effective carbon dioxide reforming was achieved with Rh and Ru metal catalysts supported on γ-Al$_2$O$_3$. With 0.5% Rh/Al$_2$O$_3$ at a GHSV (gas hourly space velocity) = 103 × 10^{-3} h^{-1} and 800°C, the conversion of CH$_4$ reached 85.5%. Also, with this catalyst, there was no carbon deposition in the temperature range 600 to 800°C. This catalyst was successfully tested in experiments in solar receivers.[3]

Carbon dioxide reforming of methane was demonstrated in a solar furnace, using an Engelhard catalyst (0.5% Rh on alumina). The product gas temperature was maintained at 750 to 800°C. With a power input of up to 6.4 kW, the maximum methane conversion reached 84%.[4,5] A similar carbon dioxide reforming of methane was also performed with a direct absorption solar receiver/reactor, heated by a parabolic dish, applying up to 150 kW of solar power. The catalytic system was a reticulated porous alumina foam disk coated with a rhodium catalyst. Up to 97 kW of total solar power was absorbed, bringing temperatures within the absorber to a range of 550 to 1100°C. The maximum methane conversion reached almost 70%, and the chemical efficiency reached 54%.[6] In another study, ruthenium and iridium supported on Eu_2O_3 were proposed as catalysts for carbon dioxide reforming of methane to synthesis gas, in what may be a viable solar-thermal energy system. A conversion efficiency of 90% was achieved either with 1% metal loading at 673 K or with 5% loading at 1023 K.[7]

Similarly effective catalysts, not requiring expensive metals, were developed, using ultrafine single-crystal magnesium oxide (uscMgO) as support. This magnesium oxide has an extremely great surface area, and also provides an excellent support effect. With 3 mol% Ni/uscMgO as catalyst system, methane conversion at 800°C was 96%. At the lower temperature of 660°C, the relative activity of different metals supported on uscMgO was tested. The order of activity was

$$Ru > Rh > Pd > Pt > Ni$$

In a variant of the carbon dioxide reforming of methane to synthesis gas, the products were ethane, ethene, and carbon monoxide,

$$CH_4 + CO_2 \rightarrow C_2H_6 + CO + H_2O \tag{8}$$

$$CH_4 + 2CO_2 \rightarrow C_2H_4 + 2CO + 2H_2O \tag{9}$$

In both reactions, $\Delta G = 35$ kJ/mol-CH_4 at 1073 K, which are thus energetically uphill reactions. Such oxidative coupling of methane, in which CO_2 is the oxidant, was found to be promoted by PbO/MgO and BaO/CaO. The reactions were performed with a $CH_4/CO_2/O_2$ (100:100:1 by volume) gas mixture. The small amount of oxygen included was designed to make the total process thermodynamically favorable. With 50% PbO/MgO and 10% BaO/CaO as catalysts, the yields of the C_2 hydrocarbons were 4.4 and 3.5%, respectively.[9,10]

The catalytic reduction of carbon dioxide by methane or higher hydrocarbons has the advantage of not requiring the energy- and cost-intensive production of hydrogen. The carbon dioxide reforming of propane and propene is particularly attractive. The catalytic reduction of carbon dioxide accompanying the dehydrogenation of propane on the oxide catalysts Cr_2O_3, ZnO, and Ga_2O_3 was tested by a pulse reaction technique,[11,12]

$$CO_2 + C_3H_8 \rightarrow CO + C_3H_6 + H_2O \tag{10}$$

There occurred considerable decomposition of propane on the chromium oxide and zinc oxide catalysts. Also, the production of propene was retarded by the presence of carbon dioxide. Only on gallium oxide were propane and carbon

$H = U + P V$

$H = G + ST$

dioxide converted to propene and carbon monoxide without the formation of by-products.[11,12]

In the aromatization of lower alkanes such as propane,

$$CO_2 + C_3H_8 \rightarrow CO + aromatics + H_2O \tag{11}$$

LPG fractions were selectively converted by the Cyclar process into petroleum-grade BTX aromatics. This process became commercial in 1990 with the start up of the first unit at the Grangemouth refinery in Scotland.[13] A typical catalyst for this aromatization of propane is Ga^{3+}-loaded ZSM-5, which enabled high selectivity to aromatics, as well as the total conversion of propane. The main products were C_6 to C_8 aromatics.[11,12,14]

Carbon dioxide reaction to carbon monoxide on Zn-loaded HZSM-5, with accompanying aromatization of propane, was studied in detail. The Zn/HZSM-5 catalyst was obtained by refluxing an aqueous solution of $Zn(NO_3)_2$ with HZSM-5 (Si/Al ratio 1:46.6). At a reaction temperature of 823 K, the C_3H_8 conversion was 68%, and organic products were aromatics (51% yield), $CH_4 + C_2H_6$ (25%), $C_2H_4 + C_3H_6$ (22%), and C_4^+ (2%). The presence of carbon dioxide suppressed the formation of coke, which occurred at such high temperatures in the aromatization of propane alone on the same catalyst.[15]

CATALYTIC HYDROGENATION

The catalytic hydrogenation of carbon dioxide to water and carbon monoxide, which is the reverse of the Water-Gas-Shift-Reaction, is an endothermic reaction,

$$CO_2(g) + H_2(g) = CO(g) + H_2O(g) \tag{12}$$

with enthalpy and free-energy changes (25°C) of $\Delta H = 41.17$ kJ/mol (9.84 kcal/mol), and $\Delta G = 28.64$ kJ/mol (6.845 kcal/mol). On the other hand, the methanation of carbon dioxide (the Sabatier reaction),

$$CO_2(g) + 4H_2(g) = CH_4(g) + 2H_2O(g) \tag{13}$$

is strongly exothermic, with $\Delta H = -164.91$ kJ/mol (-39.42 kcal/mol) and $\Delta G = -113.6$ kJ/mol (-27.14 kcal/mol) at 25°C.

To Elementary Carbon

Fixation of carbon dioxide to both elementary carbon and carbon monoxide was achieved during catalytic hydrogenation at atmospheric pressure and at temperatures from 573 to 973 K. Active catalysts for production of both carbon and CO at 973 K were in decreasing order of activity $WO_3 > Y_2O_3 > ZnO > Cr_2O_3 >$

$CO_2 + 4H_2 \rightarrow CH_4 + 2H_2O$

$CO_2 + CH_4 \rightarrow 2CO + 2H_2$

$CeO_2 > Mn_2O_3 > MgO > V_2O_5 > ZrO_2 > MoO_3$. The production of carbon was the predominant reaction below 773 K, while CO production became the favored reaction above 873 K.[16,17] This is understood on the basis of thermodynamic considerations, as the equilibrium,

$$C + CO_2 = 2CO$$

is shifted to the right with rising temperatures.[18,19] Using WO_3 as catalyst, at 973 K, the conversion of CO_2 into C and CO reached 27.6 and 42.3%, respectively. In order to enhance the selectivity for carbon production, mixed oxide catalysts were tested. With some mixed oxide materials, such as $LaFeO_3$, $LaMnO_3$, $LaMn_{0.9}Cu_{0.1}O_3$, and $PbWO_4$, the selectivity for elementary carbon formation was highest, and no CO was produced. The deposited carbon could easily be removed from the catalysts by exfoliation, thus providing an interesting method of carbon fixation.[16]

An alternative reagent for the efficiency decomposition of carbon dioxide into carbon is oxygen-deficient (H_2-reduced) magnetite. This cation-excess reagent was prepared by flowing H_2 gas through magnetite powder at 290 to 300°C for 2 h. The magnetite retained its spinel-type structure, and converted CO_2 into carbon with 100% yield at 300°C.[20,21] The elementary carbon thus formed was readily hydrogenated at 300°C into methane.[22] Similar results of the hydrogenation of carbon dioxide at 300°C to produce mainly carbon black were achieved using strontium ferrite (prepared from 95 parts of FeO and 5 parts of SrO), with a conversion of up to 98%.[23]

Wüstite, which is FeO having the cubic rock salt structure (similar to NaCl), with Fe^{2+} in the octahedral interstices, was found to be a useful reagent for the decomposition of CO_2 at 300°C. The active reagent was deficient in iron, and was prepared by heating Fe(II) oxalate under nitrogen for 15 h at 650°C. The active wüstite had the composition $Fe_{0.98}O$, and when reacted with carbon dioxide at 300°C caused its transformation into elementary carbon. Carbon monoxide appeared as an intermediate for several hours, but was then also reduced to carbon. During the CO_2 decomposition reaction, most of the wüstite was transformed into magnetite, Fe_3O_4.[24]

Methanation

Interest in the methanation reaction has been stimulated by efforts to produce synthetic methane from the synthesis gas obtained in the gasification of coal.[25] The methanation reaction is usually performed on catalysts of nickel supported on alumina, at temperatures of 300 to 400°C, and is applied as the final CO_2 and CO removal step in the purification of hydrogen to be used in ammonia synthesis.[26]

Over a reduced iron oxide catalyst (prepared by treating Fe_3O_4 under H_2 at 400 or 500°C), mixtures of CO_2 and H_2 in different proportions were passed at atmospheric pressure. The products included CO, CH_4, C_2H_4, C_2H_6, and C_3 compounds. Optimal production of hydrocarbons was obtained at a reaction temperature of 400°C, with a CO_2/H_2 ratio of 1/4 and a space velocity of 3000 cm³ g⁻¹ h⁻¹.[27] The distribution of products can be interpreted by a mechanism in which the primary step is reduction to CO,

$$CO_2 + H_2 \rightarrow CO_{ads} \tag{14}$$

followed by a Fischer-Tropsch reaction,

$$nCO_{ads} + mH_2 \rightarrow C_nH_{2m} \tag{15}$$

High selectivity in the hydrogenation of carbon dioxide to methane was reported in some Co/Cu/K catalysts. A catalyst prepared from Co_3O_4 alone had a selectivity of 97.7% for CH_4 formation at the low temperature of 424 K. The presence of K in the catalyst matrix favored the production of CO. An increase in the ratio Cu/Co in the catalyst resulted in an enhancement in the selectivity to C_2H_6.[28]

A redox catalyst of highly dispersed sub-stoichiometric ruthenium oxide supported on titanium oxide powder was effective for the methanation of carbon dioxide at ambient temperature. The catalyst contained 5% RuO_2 (3.8% Ru) in 20-Å clusters on titanium dioxide (Degussa P25, about 80% anatase, BET specific surface 55 m² g⁻¹). In the dark, the rate of the methanation of carbon dioxide was very low at 25°C, but increased rapidly with rising temperature. Illumination with a solar simulator (at 80 mW cm⁻¹) caused a very marked enhancement in the rate of methanation, which was initially ascribed to band-gap excitation of the titanium dioxide support.[29] However, subsequent work indicated that the effect of illumination could also be explained by local heating of the catalyst material, and was not an intrinsic photochemical step.[30,31] In a variant of this reaction, a NiO/kieselguhr catalyst was preirradiated with a low-pressure mercury lamp. Carbon dioxide was then made to flow over it in the dark, producing preferentially methane, as well as small yields of ethane and ethene.[32] Ru/TiO_2 (3.8% Ru) was more active as a methanation catalyst by two orders of magnitude than Ru/Al_2O_3 (3.8% Ru). The reaction rate on Ru/TiO_2 was independent of the CO_2 concentration and was approximately one half order in H_2 concentration. An FTIR spectroscopic study revealed the formation of surface-adsorbed CO_{ads} and $HCOO^-_{ads}$ intermediates. The CO_{ads} was presumably formed by the reverse water-gas shift reaction. These species were proposed to be intermediates toward CH_4 production.[33]

Using a CO_2-H_2 gas mixture flowing over TiO_2 at atmospheric pressure, the rate of production of CO and CH_4 was measured at temperatures up to 1000°C. At 900°C, the rates of formation of CO and CH_4 reached 300 and 100 μmol min⁻¹, at a space velocity (gas flow rate/reactor volume) of 16 min⁻¹ as shown in Figure 1.[34]

The mechanism of methanation of carbon dioxide was clarified by rate measurements over the following alumina supported catalysts: Ni (17%), Ru (5%), Pt (10%), and Ir (6%), all on alumina. From a kinetic analysis of the rates of

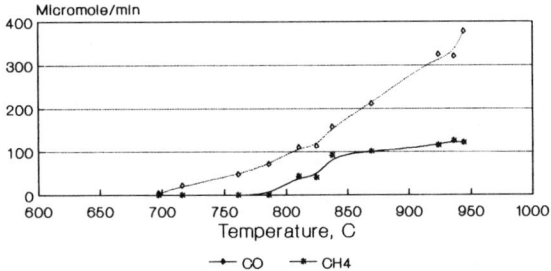

Fig. 1 Hydrogenation of carbon dioxide on TiO_2.

evolution of the products CO and CH_4, the existence of two different pathways was indicated: a parallel reaction mechanism was observed with Ni-alumina and Ru-alumina. On the other hand, with Pt-alumina and Ir-alumina, a consecutive reaction mechanism was predominant,[35]

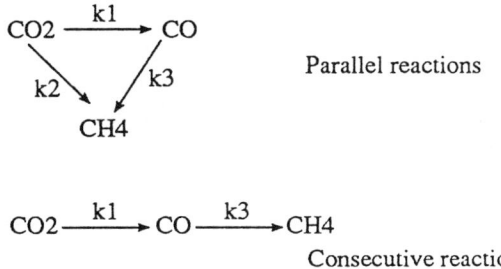

Rapid methanation of carbon dioxide with hydrogen was achieved using a composite catalyst, Ni-La_2O_3-Ru (4.3 wt%, 2.5 wt%, 0.7 wt%) dispersed on a spherical silica support having a meso-macro bimodal pore structure. Total CO_2 conversion was obtained, with 100% selectivity for methane. The space-time yield of methane reached 500 mol l^{-1} h^{-1}, at a temperature range around 270°C. The function of the La_2O_3 part was to enhance the CO_2 adsorption capacity, while the presence of the Ru enhanced the absorption of hydrogen and presumably facilitated the hydrogen spillover.[36,37]

The carbides Mo_2C, Fe_3C, and WC were found to be quite effective catalysts for the hydrogenation of carbon dioxide. Thus, using copper-promoted molybdenum carbide, Cu/Mo_2C, with a gas mixture of Ar/H_2/CO_2 = 10.1/22.7/67.2 at 220°C, the CO_2 conversion was 4%, and the selectivity to methanol, dimethyl ether, methane, ethane, and CO was 31.5, 1.4, 13.5, 3.0, and 48.5%.[38]

A glassy metal alloy, $Ni_{64}Zr_{36}$, was used as catalyst precursor. Under CO_2 hydrogenation conditions (at 493 K), the amorphous alloy was transformed to a microporous solid, with metallic nickel particles contained in a ZrO_2 matrix. Methane was the almost exclusive CO_2 reduction product, along with traces of ethane.[39]

The exothermic methanation of carbon dioxide with hydrogen was used to close the loop of the above-mentioned solar furnace-driven carbon dioxide reforming. With the same Rh/alumina catalyst as used for the reforming, conversion of over 80% was achieved in the methanation reaction.[5]

Since most of the carbon on Earth ($>99.9\%$) exists as carbonates such as limestone — $CaCO_3$, it may sometimes be useful also to reduce solid carbonates. Alkaline earth carbonates, when heated under hydrogen, underwent degradation at reaction temperatures which were lower by at least 150 K compared with the degradation under inert or oxidizing atmospheres. Magnesite decomposed in a reducing atmosphere to MgO, as well as to even amounts of CO_2 and CO. From calcite, the main gaseous product was CO, with the $CO:CO_2$ ratio 10:1. With mixed alkaline earth/transition metal carbonates (prepared by co-precipitation from the nitrate salt solutions with Na_2CO_3) under H_2, the decomposition temperature was even further lowered. Thus, with Ca-Ni carbonates (10% Ni) under H_2, the decrease was 400 K compared with the decomposition temperature of pure $CaCO_3$ under a nonreducing atmosphere. From Ca-Co, Ca-Ni, and Mg-Ni (10% transition metal) carbonates, the predominant gaseous product was CH_4, with small amounts or traces of CO[40]

By hydrogenation at 200 to 250°C, $NiCO_3$ and $CoCO_3$ produced methane, even without any added catalysts. $CaCO_3$ and $BaCO_3$ were hydrogenated to CH_4 in the presence of Ni or Co catalyst (0.01 g catalyst/1 g carbonate) at 400°C. Very rapid hydrogenation occurred with $4MgCO_3 \cdot Mg(OH)_2 \cdot 5H_2O$ at 300°C in the presence of a catalyst of ultrafine Ni powder, reaching a production rate of 3.2 mmol CH_4 h^{-1} (g substrate)$^{-1}$.[41]

Methanol Synthesis

The catalytic hydrogenation of mixtures of carbon monoxide and carbon dioxide to produce methanol is one of the basic processes in the huge petrochemical industry.[42] Together with carbon monoxide, carbon dioxide can also serve as the carbon source in the Fischer-Tropsch synthesis of mixtures of various aliphatic and aromatic hydrocarbons, aldehydes, ketones, and carboxylic acids. Selective hydrogenation to yield mainly methanol has been achieved with catalysts based on copper, chromium, zinc, and palladium, while methanation of carbon dioxide is the preferred reaction on catalysts of nickel supported on alumina or rhodium supported on titania.[43]

Comparative studies, at atmospheric pressure, of hydrogenation on Cu-Zn catalysts supported on Al_2O_3 indicated that the conversion to methanol, according to

$$CO_2 + 3H_2 = CH_3OH + H_2O \qquad (17)$$

occurred at lower temperatures and with higher yields than that of the mixture of $CO + H_2$.[44] Cu/ZnO catalysts were activated by reduction in a H_2 stream at 300°C, and applied to the hydrogenation of CO_2 and CO to methanol at 225°C. The product selectivity was found to depend on the contact time of the reactants over the catalyst. Prolonged contact times favored the formation of homologous hydrocarbon products.[45]

The mechanism of the catalytic hydrogenation of carbon dioxide over a variety of metal/zirconia catalysts has been studied in detail. On Pd/ZrO_x catalysts, prepared by activation of Pd_1Zr_2 alloys over CO_2-H_2 mixtures, at a reaction temperature of 463 K, the reactant conversion was 8.5%. Methanol, methane, and CO were produced with selectivities of 43, 18, and 39%.[46] In a diffuse reflectance FTIR study of adsorbed species on the activated Pd/ZrO_2 catalyst, the disappearance of surface formate was correlated with the appearance of gas phase methane.[47] While Pd/ZrO_2 catalysts favored selectivity to methane, Cu/ZrO_2 catalysts promoted selectivity to methanol. With catalysts obtained from an amorphous $Cu_{70}Zr_{30}$ precursor, diffuse reflectance FTIR spectroscopy measurements indicated a mechanism involving: (1) rapid adsorption of CO_2 followed by its reduction to surface formate, which was further reduced to methane. (2) CO_2 on the surface may be hydrogenated to CO by the reverse water gas shift reaction. Adsorbed CO in the presence of hydrogen produced π-bonded formaldehyde, which was reduced to methylate, and finally to methanol.[48-50] With amorphous $Au_{25}Zr_{75}$ alloy as precursor, activated *in situ* under CO_2 hydrogenation conditions, the gold was reduced to metallic particles of 8.5-nm mean size, while the zirconia was oxidized to ZrO_2. With this catalyst, the main products of CO_2 hydrogenation were methanol and carbon monoxide. However, on this catalyst, the selectivity for methanol is worse than on the Cu/ZrO_2 catalyst. Diffuse reflectance FTIR spectroscopic investigation of adsorbed species during the reaction on the Au/ZrO_2 catalyst indicated the appearance of two types of formate species, with doublets of bands at 1580/1380 cm^{-1} (type I) and at 1600/1360 cm^{-1} (type II).[51]

Rhenium catalysts supported on ZrO_2 or Nb_2O_5 were found to be effective for methanol synthesis from $CO_2 + H_2$. Re-ZrO_2 gave excellent reactivity and high selectivity (73%) for methanol formation at 160°C and 10 atm pressure. With Re-Nb_2O_5, selectivity for methanol was also good (>50%), but the reactivity was lower. Other CO_2 reduction products were CO, CH_4, and dimethyl ether.[52]

The effect of the supports MgO, Al_2O_3, SiO_2, TiO_2, ZrO_2, and Nb_2O_5 was tested for the $CO_2 + H_2$ synthesis of methanol on ZnO catalysts. Highest activity and selectivity for methanol production was with the system ZnO/ZrO_2 (1:9), at 360°C. With this system, methanol synthesis was more selective using $CO_2 + H_2$ than using $CO + H_2$. Acidic supports, such as Al_2O_3, TiO_2, and SiO_2 favored the dehydration of methanol to dimethyl ether. Therefore, methanol production was more selective with less acidic or even basic supports, such as ZnO, MgO, and ZrO_2.[53]

Copper catalysts supported on zirconia are of particular interest because of their high selectivity and activity for methanol synthesis from CO_2 and H_2. In a careful study of several preparation methods for Cu/ZrO_2 catalysts, the procedures tested included impregnation, ion exchange, deposition-precipitation, co-precipitation, and simultaneous precipitation and reduction in the presence of reducing agents. The formation of metallic copper, Cu^o, on the catalysts was found to be essential for good methanol productivity. High activity and selectivity for methanol synthesis from carbon dioxide and hydrogen was observed with catalyst systems providing large interfacial areas between the CuO and the ZrO_2 components. The reaction mechanism probably involved a common surface intermediate for the two parallel reaction pathways leading to the production of methanol and of gaseous CO.[50]

When only CO_2 was used as the carbon source, the rate of methanol synthesis on the model catalysts of Cu/ZnO was slower than with CO_2-CO mixtures.[54,55] With the commercial catalyst of $Cu/Zn/Al_2O_3$ (ICI), under industrial conditions (250°C, 40 to 50 atm), a GHSV of 10,000 to 120,000 h^{-1} resulted in the direct hydrogenation of CO_2 to methanol. Isotope-tracer experiments proved the absence of a carbon-containing surface intermediate which was common to methanol synthesis and to the water-gas shift reaction. This was indicated by the hydrogenation of $^{12}CO/^{14}CO_2$ mixtures.[56] At low conversions, the ^{14}C label appeared in the methanol produced.[57,58] On model $Cu-ZnO-Al_2O_3$ catalysts, CO_2 treatment resulted primarily in the generation of adsorbed carbonate and hydrogeno carbonate species. Hydrogenation of these carbonates resulted in the conversion to adsorbed formate species, as indicated by FTIR measurements.[59,60] The proposed mechanism involves this adsorbed formate intermediate,

$$H_2 \rightarrow H^- + H^+ \tag{18}$$

$$CO_2 + H^- \rightarrow HCOO^- \tag{19}$$

$$HCOO^- + 2H_2 \rightarrow CH_3O^- + H_2O \tag{20}$$

The methoxide intermediate is then converted to methanol by hydrolysis or hydrogenolysis.[55,61]

The kinetics of methanol synthesis at various H_2, CO, and CO_2 compositions was measured over Cu/ZnO (Cu:Zn = 30:70 atomic ratio) and $Cu/ZnO/Al_2O_3$ (Cu:Zn:Al = 60:30:10 atomic ratio). The ternary catalyst was more active for methanol production, but also generated methane as a minor by-product (5 to 10% of the methanol yield). The kinetic data proved that methanol was formed from either CO or CO_2.[62]

The mechanism of methanol synthesis on the highly effective Cu/ZnO catalyst (Cu/Zn = 3:7) was studied by diffuse reflectance FTIR spectroscopy and by temperature-programmed desorption. The catalyst was prereduced under H_2, resulting in the formation of metallic copper. When $CO_2 + H_2$ (1:9) was fed over the catalyst, methanol synthesis occurred, in parallel with the reverse water gas shift reaction (producing CO and H_2O). The adsorbed species on the catalyst identified after the reaction included zinc methoxide $CH_3O\text{-}Zn$, bidentate zinc formate, and copper formate HCOO-Cu. The zinc methoxide species, with IR absorption maxima at 2930 cm^{-1} (asymmetrical CH_3 stretching), 2825 cm^{-1} (symmetrical CH_3 stretching), and 1060 cm^{-1} (CO stretching), was shown to be the precursor for methanol synthesis. Methanol was released from the catalyst by hydrolysis with water — which was formed by the reverse water gas shift reaction.[63]

The outstanding importance of the catalyst support was revealed in a comparative study of the catalyst-support systems Ni(3.8)-Mo(7.0)/Al_2O_3 and Ni(3.0)-Mo(7.9)/ZnO. In experiments at a H_2/CO_2 ratio of 2, at 553 K and a pressure of 3 MPa, the selectivities after 7 to 8 h of reaction were with the Al_2O_3-supported catalyst: methanol, 9%, methane, 50%, and CO, 38%. With the ZnO-supported catalyst, the selectivities were methanol, 25%, methane, 1%, and CO, 73%. From the time course of the evolution of the products, the authors concluded that methanol production seemed to be mainly by the direct hydrogenation of CO_2, although some methanol may be derived from the hydrogenation of CO. The results showed that methanol formation was favored on the ZnO support.[64]

An improved copper catalyst for the hydrogenation of carbon dioxide to methanol was developed by achieving good dispersion of the metal in a stable pyrochlore matrix. The pyrochlore support $La_2Zr_2O_7$, which has good thermal and chemical stability, was prepared by the reactions,

$$CuO + La_2O_3 \rightarrow CuLa_2O_4 \qquad (21)$$

$$CuLa_2O_4 + 2ZrO_2 \rightarrow La_2Zr_2O_7 + CuO \qquad (22)$$

At high pressure conditions (6 MPa), the methanol production rate reached 530 g methanol kg.cat^{-1} h^{-1}. Using this catalyst, high selectivity for methanol formation (vs. the competing water-gas-shift reaction) was obtained by operating at a low reaction temperature, high reaction pressure, high H_2 partial pressure, and high H_2/CO_2 ratio.[61]

The effect of the additives ZrO_2, MgO, Al_2O_3, or Cr_2O_3 on the activity of CuO-ZnO catalysts for methanol synthesis from CO_2 and H_2 at 13 atm pressure was studied in a flow reactor. With the catalyst additives Al_2O_3, ZrO_2, and MgO, the activity and selectivity for methanol was higher than with CuO-ZnO alone. With a CuO-ZnO-ZrO_2 catalyst (molar ratio 42/47/11), the highest yield (17.4%) of methanol was at 220°C, and the highest selectivity (94.8%) for methanol was at 160°C.[65]

At the very low reactant pressure of 5 atm, CO_2 and H_2 were converted to methanol on a Re catalyst supported on CeO_2, ZrO_2, and La_2O_3. The highest selectivity for methanol formation (76.7%) was observed over Re/CeO_2 at a reaction temperature of only 160°C.[66] In order to distinguish between the direct hydrogenation of CO_2 to methanol, and the alternative mechanism of CO formation by the reverse water-gas shift reaction, experiments were made on mixtures of

^{13}CO, CO_2, and H_2. The extent of ^{13}C incorporation into the methanol indicated the mechanism. Over CuO-ZnO catalysts, methanol was produced almost exclusively by direct hydrogenation of CO_2. Over Re/ZrO$_2$ and Re/CeO$_2$, both routes were operative. The reaction involved surface formate and adsorbed formaldehyde as intermediates in methanol synthesis.[67]

A highly active catalyst for methanol synthesis from carbon dioxide and hydrogen was prepared by an "intrinsic uniform gelation method", in which the concentrated mixed nitrate aqueous solutions of the catalyst metals was treated with ammonia gas, to form a gel, which was then dried and calcined. With CuO-ZnO-Cr$_2$O$_3$-Al$_2$O$_3$ (25.0-41.5-1.2-32.3 wt%), modified by addition of 6% Pd to enhance hydrogen spillover, the space-time yield of methanol amounted to the very high value of 356 g l^{-1} h^{-1}. Tests at 250°C indicated that increases in the reaction pressure resulted in improvements in both the yield and the selectivity for methanol. An interesting further development was to connect the outlet of the above Pd-CuO-ZnO-Cr$_2$O$_3$-Al$_2$O$_3$ reactor, releasing methanol as well as unreacted excess hydrogen, into a H-Fe-silicate reactor connected in series, which converted the methanol into a light gasoline. Gasoline was obtained with 45% selectivity in a simple one-pass operation. The selectivity to gasoline could be increased by injection of a small amount of propylene into the inlet of the second reactor. Propylene enhanced the autocatalytic conversion of light olefins to gasoline.[37,68,69]

An alternative process for the hydrogenation of carbon dioxide used novel hybrid catalysts, which combine the formation of methanol and its conversion to hydrocarbons. These catalysts were prepared by physical mixing of the methanol synthesis catalyst, preferably oxides of Cu-Zn-Cr (2:2:1 or 3:3:1) containing hexavalent chromium, with zeolites (such as the synthetic zeolite JRC-Z-HY4.8 with Si/Al = 4.8). Typical conditions of conversion were 400°C, 50 kg cm^{-2}, SV = 3 l h^{-1} g^{-1}, H$_2$/CO$_2$ = 3. The yields of hydrocarbons (mainly C$_2$ to C$_4$) reached 8.6%.[70]

Methanol and the higher alcohols ethanol, 1-propanol, and 2-methyl-2-propanol were produced over Cu/ZnO/Cr$_2$O$_3$ catalysts, with an additive of 0.5% K$_2$CO$_3$ as promoter, using CO$_2$, CO, and H$_2$ feed gases at 10 MPa and 285 to 315°C. It was suggested that CO$_2$ participated directly in higher alcohol synthesis on copper sites.[7]

The synthesis of methanol from CO$_2$ and H$_2$ over Cu(111) and Pd(111) surfaces was analyzed using the "bond order conservation-Morse Potential" (BOC-MP) approach. In this semiempirical analysis, the probability of the intermediate reaction mechanisms can be evaluated, based on calculations of the heats of chemisorption of all adsorbed species, and of the activation barriers for all elementary reactions assumed to be involved in the synthesis of methanol. In the hydrogenation of CO$_2$ on Cu, the activation barrier for intermediate formate production was shown to be significantly lower than that for the disporportionation into CO$_s$ and OH$_s$. On the other hand, on Pd catalyst, the activation barriers for both reactions were found to be essentially equal. Thus, the formate intermediate route is considered the preferred mechanism for CO$_2$ hydrogenation to methanol on copper.[72]

High methanol selectivity requires high gas pressures and low rea[...] peratures. Optimal conditions for methanol synthesis on a CuO-ZnO-Al$_2$ catalyst were obtained at the condition of 7-mPa pressure, 250°C tem[...] space velocity GHSV = 1800 h^{-1} and H$_2$/CO$_2$ = 3/1, resulting in 79% se[...] for methanol production, 20% yield, and 25% CO$_2$ conversion.[73]

TiO$_2$ as a support in methanol synthesis was found to suppress the reverse water gas shift reaction. An optimal methanol synthesis catalyst had the composition CuO-ZnO-TiO$_2$ (30:30:40 by weight). Using the gas mixture H$_2$/CO$_2$ = 4/1, at 513 K, the conversion was 15.4%, the selectivity to methanol 21.9%, and the yield of methanol 3.4%.[74]

Formate Synthesis

Formic acid is an important chemical, useful as a pickling agent, as reducing agent, in the manufacture of animal silage, and as an intermediate in the production of oxalic acid, formate esters, and amides such as dimethylformamide.[75,76]

The selective catalytic hydrogenation of bicarbonate ions in aqueous solutions to formate ions was achieved using Pd supported on carbon or γ-Al$_2$O$_3$ as catalysts,

$$H_2 + HCO_3^- = HCOO^- + H_2O \tag{23}$$

The reaction did not go to completion; rather, an equilibrium was reached. Thus, with initially 1.0 M NaHCO$_3$/1.7 atm H$_2$, the HCOO$^-$ concentration reached about 0.54 M.[77] The rate of hydrogenation increased with the hydrogen gas pressure, until a plateau was reached, in accordance with the Langmuir isotherm law. Because of the common ion effect, the maximal concentrations of formate which were obtained at 6 atm. H$_2$ and 35°C were 2.5 M for sodium formate and 5.8 M for potassium formate. The addition of CO$_2$ gas to the bicarbonate reactant did not result in an appreciable increase in the rate of hydrogenation. Therefore, presumably, the bicarbonate ion was the active species undergoing hydrogenation.[78]

The hydrogenation of carbon dioxide to formic acid may be accomplished also in homogeneous solutions, using transition metal catalysts. With [Rh(diene)L$_3$]Z (L = R$_3$P) complexes as catalysts, in tetrahydrofuran solution at room temperature, formic acid was generated with a turnover number of 34 within 3 d. The turnover number could be enhanced by pretreatment of the catalyst with molecular hydrogen, and by including some water (0.4% v/v) in the reaction mixture, thus achieving a turnover number of 128 within 2 d. With norbornadiene (NBD) as the diene ligand, such as with [Rh(NBD)(PMe$_2$Ph)$_3$]BF$_4$ in the above precatalyst, the hydrogenation of CO$_2$ to HCOOH could be performed in the THF solution, with CO$_2$ and H$_2$ each pressurized to 700 to 750 psi. The addition of H$_2$ to the precatalysts generated rhodium dihydride complexes, which presumably were the active catalysts.[75,76]

handwritten at top: $2H_2 + HCOOCH_3 \rightarrow$
$2CH_3OH$

Direct formation of formic acid from carbon dioxide and dihydrogen was also achieved using the homogeneous $[\{Rh(cod)Cl\}_2]$-$Ph_2P(CH_2)_4PPh_2$ catalyst system, in which (cod) = cycloocta-1,5-diene. This catalyst was prepared *in situ* from $\{Rh(cod)Cl\}_2$ and $Ph_2P(CH_2)_4PPh_2$ in DMSO solution containing triethylamine. With a total initial $CO_2 + H_2$ pressure of 40 atm, up to 1150 mol of formic acid per mole of rhodium were formed within 24 h at room temperature.[79]

An interesting modification of the synthesis gas conversion to formic acid is the hydrocondensation of CO_2 and H_2 in the presence of methanol and anionic ruthenium clusters to produce methyl formate,

$$CO_2 + H_2 + CH_3OH \rightarrow HCOOCH_3 + H_2O \qquad (24)$$

Methyl formate is useful as it can be catalytically isomerized to acetic acid. The production of methyl formate was effectively catalyzed by the complex $HRu_3(CO)_{11}^-$, at a temperature of 125°C and pressures of 250 psi CO_2 and 250 psi H_2. The catalyst turnover reached up to 4 per 24 h. During the reaction, the $HRu_3(CO)_{11}^-$ complex was converted into a tetraruthenium cluster, $H_3Ru_4(CO)_{12}^-$, which was even slightly more active than the original catalyst.[80] In a study of the mechanism of the reaction of the $HRu_3(CO)_{11}^-$ with carbon dioxide, an insertion product was identified. This carboxylation occurred slowly at 60 psi pressure, but more rapidly at 400 psi pressure of carbon dioxide.[81]

$$HRu_3(CO)_{11}^- + CO_2 = HCO_2Ru_3(CO)_{10}^- + CO \qquad (25)$$

Reverse Water Gas Shift Reaction

The reverse water gas shift reaction is energetically an uphill reaction, thus storing energy,

$$CO_2(g) + H_2(g) = CO(g) + H_2O(g) \quad \Delta H = 9.84 \text{ kcal mol}^{-1} \qquad (26)$$

Highly effective catalysts for the reverse water gas shift reaction were developed using finely divided molybdenum sulfide on various support materials such as TiO_2. The catalyst was prepared by impregnating the support material with an ammoniacal solution of ammonium tetrathiomolybdate, followed by drying and calcining at 350°C. The activity of the supported MoS_2 at 400°C for the hydrogenation of CO_2 to CO depended on the support in the order $TiO_2 > Al_2O_3 > ZrO_2 > SiO_2 > CeO_2 > MnO_2$. The selectivity to CO production was more than 99.5%.[82]

In a mechanistic study of the reverse water gas shift reaction, a clean Cu(110) single crystal model catalyst was tested at temperatures of 573 to 723 K. The kinetic data could best be interpreted by a "surface redox" or "oxygen adatom" mechanism for both the forward and the reverse water gas shift reactions. For the reverse water gas shift reaction, the proposed rate determining step is dissociative adsorption of CO_2, with the formation of adsorbed CO (CO_a) and oxygen atoms (O_a),[83]

$$CO_2 \leftrightarrow CO_a + O_a \tag{27}$$

$$CO_a \leftrightarrow CO \tag{28}$$

$$H_2 \leftrightarrow 2H_a \tag{29}$$

$$H_a + O_a \leftrightarrow OH_a \tag{30}$$

$$H_a + OH_a \leftrightarrow H_2O_a \tag{31}$$

$$H_2O_a \leftrightarrow H_2O \tag{32}$$

A homogeneous solution phase reaction similar to the reverse water-gas shift reaction was observed with some Rh complexes. Rhodium(I) complexes $Rh[PL_3]_3$, such as $Rh[P(i\text{-}Pr)_3]_3$, reacted with CO_2 in the presence of water to form dihydro bicarbonate complexes, such as $Rh_2(O_2COH)(i\text{-}Pr)_2$. These dihydro bicarbonate complexes reduced CO_2 to form Rh(I)carbonyl-bicarbonate complexes, in an analogue of the reverse water gas shift reaction,[84]

MICROWAVE INDUCED REDUCTION

The endothermic carbon dioxide reforming reaction,

$$CO_2 + CH_4 \rightarrow 2CO + 2H_2 \tag{33}$$

with $\Delta H^\circ = +247.27$ kJ mol^{-1} was activated by a plasma discharge, using microwave radiation at 2450 MHz, with an incident power of 80 W. The reaction was performed in a closed circulation system, through a Pyrex glass tube, at a gas pressure of 5 torr. The microwave discharge on methane alone gave acetylene as the main product. On the other hand, with methane + carbon dioxide (1:1), after 4 min, there was 96.5% conversion of CH_4, and the products were CO (46%), H_2 (40%), $^2C + ^3C$ hydrocarbons (2%).[85]

Microwave-induced catalysis was applied to promote the reaction of carbon dioxide and water on a Ni/NiO catalyst (which had been activated at 440°C under H_2 + He for 5 to 6 h). This technique enables the selective absorption of microwave energy and rapid surface heating of the catalyst, with very small energy absorption by the reactants or the support material. The energy source was a 2.45-GHz magnetron providing 168-ms pulses, with 20-s dark times, with an average incident power of 2.2 kW. After a total irradiation time of 29.5 s, using a H_2O:CO_2 = 1:2.5 ratio, the product composition was CH_4 (55.1%), C_2H_6 (0.3%), CH_3OH (5.5%), acetone (4.7%), C_3 alcohols (5.8%), and C_4 alcohols (28.4%). No CO_2 reduction occurred in the absence of the catalyst or the microwave radiation. It was proposed that one of the initiating events in the reaction was the microwave-induced decomposition of water, forming H atoms and OH radicals on the catalyst surface.[86]

SONOLYSIS

Ultrasound-initiated redox reactions in aqueous solutions occur only in the presence of monoatomic or diatomic gases. Monoatomic gases such as argon are effective for sonolysis because of their low specific heat, resulting in high temperature rises in the small gas bubbles formed. These gas bubbles are produced by the cavitation due to the ultrasound. During the very rapid adiabatic compression, temperatures of several thousand degrees Kelvin and extreme pressure fluctuations cause water dissociation, forming free radicals, such as $\cdot H$ and $\cdot OH$,[87]

$$H_2O \rightarrow \cdot H + \cdot OH \tag{34}$$

Pure CO_2 in water did not undergo sonolysis, while, with pure water under argon, the only products were H_2 and H_2O_2. With carbon dioxide in water, in the presence of argon, CO was produced, as well as traces of HCOOH; the yield of H_2 decreased and the yield of H_2O_2 increased. Even with only 0.5% CO_2 in the argon gas mixture did the H_2 yield decrease to one half. Irradiations were made with a 300-kHz quartz oscillator, providing a high-frequency power of 3.5 W cm^{-2}. Carboxylation reactions were observed during the irradiation of 0.05 M ethanol in an Ar-CO_2 atmosphere, resulting in the production of formic, acetic, and lactic acids, while ethanol was also decomposed to acetaldehyde and formaldehyde. The free radical reactions which were proposed to account for the sonolysis products differed from those occurring in photolytic and radiolytic processes in liquid media, and were more typical of those occurring in flames. The very high local concentrations of radicals favored radical-radical rather than radical-molecule reactions. Primary reaction steps in the sonolysis of carbon dioxide in water were proposed to be both protonation and direct dissociation,[87]

$$CO_2 + \cdot H \rightarrow \cdot COOH \tag{35}$$

$$CO_2 \rightarrow CO + O \tag{36}$$

GASIFICATION OF CARBON

The gasification of carbonaceous materials such as coal by carbon dioxide is an attractive method for both the disposal of waste carbon dioxide and production of carbon monoxide,

$$CO_2 + C = 2CO \tag{37}$$

This reaction was found to be catalyzed by alkali metal ions. In the presence of 0.9 wt% potassium carbonate, the above reaction became very rapid at 700°C, with an apparent activation energy E_a of 159 kJ mol^{-1}, while in the absence of potassium carbonate the reaction was just measurable at 700°C, with an apparent activation energy of 360 kJ mol^{-1}.[88]

Molybdenum oxide, MoO_2, was used as a catalyst for the carbon dioxide gasification of activated carbons and chars in the temperature range 773 to 823 K. The apparent activation energy E_a and the preexponential factor $A = 1nR_0$ were obtained from the Arrhenius equation, giving values of 41 to 70 kJ mol^{-1} and 7 to 11, respectively. The addition of sulfur to a carbon sample (C/S = 1/5 by weight) resulted in a decrease in the rate of gasification to one third, and an increase in E_a from 41 to 75 kJ mol^{-1}. Thus, sulfur caused a partial poisoning of the MoO_2 catalyst.[89]

Sodium lignosulfonate was found to be an effective catalyst for the gasification of active charcoal by carbon dioxide. This catalysis involved decreases in both the kinetic parameters E_a and A. On the other hand, $Fe(NO_3)_3$ as catalyst for charcoal gasification caused increases in these kinetic parameters.[90]

The gasification by CO_2 of single-crystal graphite (a model of the gasification of coal) was compared with the gasification by steam. The crystals of natural graphite were cleaved and then etched by exposure to the gases at 600°C and 23-mm pressure. Surface vacancies were thus expanded, creating one atomic layer deep pits. The edges of pits thus formed on the graphite crystal face were decorated with gold nuclei, and were examined by transmission electron microscopy (TEM). The relative reactivities of H_2O and CO_2 were 11:1. Also, the pits formed from CO_2 were circular, while the pits formed from H_2O were hexagonal.[91]

The carbon-carbon dioxide reaction was measured on a polyvinyl coke, which had been heat pretreated at 1600°C to remove hydrogen. The technique used was the temperature-programmed desorption of quenched carbon-oxygen surface complexes. From the Arrhenius plots of the rates of gasification vs. the temperature, an apparent activation energy of 181 kJ mol^{-1} was obtained. This activation energy was equal to the apparent activation energy of the desorption step. This may indicate that the desorption reaction is the rate-limiting step.[92]

References

1. **Teuner, S. C.**, Make carbon monoxide from carbon dioxide, *Hydrocarbon Processing*, 64, 106–107, 1985.

2. **Solymosi, F., Kutsán, Gy., and Erdöhelyi, E.**, Catalytic reaction of CH_4 with CO_2 over alumina-supported Pt metals, *Catal. Lett.*, 11, 149–156, 1991.

3. **Richardson, J. T. and Paripatyadar, S. A.**, Carbon dioxide reforming of methane with supported rhodium, *Appl. Catal.*, 61, 293–309, 1990.

4. **Fraenkel, D., Levitan, R., and Levy, M.**, A solar thermochemical pipe based on the CO_2-CH_4 (1:1) system, *Int. J. Hydrogen Energy*, 11, 267–277, 1986.

5. **Levitan, R., Levy, M., Rosin, H., and Rubin, R.**, Closed-loop operation of a chemical heat pipe at the Weizmann Institute solar furnace, *Solar Energy Mater.*, 24, 464–477, 1991.

6. **Buck, R., Muir, J. F., Hogan, R. E., and Skocypec, R. D.**, Carbon dioxide reforming of methane in a solar volumetric receiver/reactor: the CAESAR project, *Solar Energy Mater.*, 24, 449–463, 1991.

7. **Perera, J. S. H. Q., Couves, J. W., Sankar, G., and Thomas, J. M.**, The catalytic activity of Ru and Ir supported on Eu_2O_3 for the reaction CO_2 + CH_4 — reversible $2H_2$ + 2CO — a viable solar-thermal energy system, *Catal. Lett.*, 11, 219–225, 1991.

8. **Matsuura, I. and Takayasu, O.**, Carbon dioxide reforming of methane using transition metal catalysts supported on magnesium oxide, Proc. Int. Symp. Chemical Fixation of Carbon Dioxide, Nagoya, Japan, Dec. 2–4, 1991, 247–252.

9. **Nishiyama, T. and Aika, K.**, Mechanism of the oxidative coupling of methane using CO_2 as an oxidant over PbO-MgO, *J. Catal.*, 122, 346–351, 1990.

10. **Aika, K. and Nishiyama, T.**, Formation of ethene and carbon monoxide from methane and carbon dioxide, Proc. Int. Symp. Chemical Fixation of Carbon Dioxide, Nagoya, Japan, Dec. 2–4, 1991, 413–418.

11. **Hattori, T., Komatsuki, M., Satsuma, A., and Murakami, Y.**, Catalytic reduction of carbon dioxide by lower alkane, *Nippon Kagaku Kaishi*, 1991, 648–650, *Chem. Abstr.*, 115, 34719z.

12. **Hattori, T., Yamauchi, S., Komatsuki, M., Satsuma, A., and Murakami, Y.**, Catalytic reduction of carbon dioxide by using lower hydrocarbons, Proc. Int. Symp. Chemical Fixation of Carbon Dioxide, Nagoya, Japan, Dec. 2–4, 1991, 419–422.

13. **Gosling, C. D., Wilcher, F. P., and Pujado, P. R.**, LPG conversion to aromatics, Proc. 69th Annu. Conv. — Gas Process. Assoc., 1990, 190–194, *Chem. Abstr.*, 114, 46184.

14. **Kitagawa, H., Sendoda, Y., and Ono, Y.**, Transformation of propane into aromatic hydrocarbons over ZSM-5 zeolites, *J. Catal.*, 101, 12–18, 1986.

15. **Hattori, T., Yamauchi, S., Satsuma, A., and Murakami, Y.**, Catalytic reduction of carbon dioxide on Zn-loaded HZSM-5 accompanying aromatization of propane, *Chem. Lett.*, 1992, 629–630.

16. **Ishihara, T., Fujita, T., Mizuhara, Y., and Takita, Y.**, Fixation of carbon dioxide to carbon by catalytic reduction over metal oxides, *Chem. Lett.*, 1991, 2237–2240.

17. **Takita, Y., Ishihara, T., and Fujita, T.**, Catalytic reduction of carbon dioxide to carbon and carbon monoxide, Proc. Int. Symp. Chemical Fixation of Carbon Dioxide, Nagoya, Japan, Dec. 2–4, 1991, 179–184.

18. **Mamantov, G.**, Molten salt electrolytes in secondary batteries, in *Materials for Advanced Batteries*, Murphy, D. W., Broadhead, J., Steele, B. C. H., Eds., Plenum Press, New York, 1980, 111–122.

19. **Deanhardt, M. L., Stern, K. H., and Kende, A.,** Thermal decomposition and reduction of carbonate ion in fluoride melts, *J. Electrochem. Soc.,* 133, 1148–1152, 1986.

20. **Tamaura, Y. and Tabata, M.,** Complete reduction of carbon dioxide to carbon using cation-excess magnetite, *Nature,* 346, 255–256, 1990.

21. **Tamaura, Y.,** CO_2 decomposition and conversion into CH_4 at 250–350°C using the H_2-reduced magnetite, Proc. Int. Symp. Chemical Fixation of Carbon Dioxide, Nagoya, Japan, Dec. 2–4, 1991, 167–172.

22. **Nishizawa, K., Tabata, M., Kodama, T., Abe, H., Yoshida, T., and Tamaura, Y.,** Methanation of CO_2 with the oxygen-deficient magnetite at 150–300°C, Proc. Int. Symp.

23. **Korenaga, T., Kaseno, S., and Takahashi, T.,** Low-temperature carbon dioxide reduction with ferrite contactor, Proc. Int. Symp. Chemical Fixation of Carbon Dioxide, Nagoya, Japan, Dec. 2–4, 1991, 381–396.

24. **Akanuma, K., Kodama, T., Tabata, M., Yoshida, T., and Tamaura, Y.,** Decomposition of CO_2 with wüstite at 300°C, Proc. Int. Symp. Chemical Fixation of Carbon Dioxide, Nagoya, Japan, Dec. 2–4, 1991, 173–178.

25. **Satterfield, C. N.,** *Heterogeneous Catalysis in Practice,* McGraw Hill, New York, 1980, 308.

26. **LeBlanc, J. R., Jr., Madhavan, S., Porter, R. E., and Kellogg, P.,** Ammonia, in *Kirk-Othmer Encyclopedia of Chemical Technology,* Vol. 2, 3rd ed., Wiley-Interscience, New York, 1978, 494.

27. **Suzuki, T., Saeki, K.-I., Mayama, Y., Hirai, T., and Hayashi, S.,** Hydrogenation of carbon dioxide over iron oxide catalyst, *React. Kinet. Catal. Lett.,* 44, 489–497, 1991.

28. **Baussart, H., Delobel, R., Le Bras, M., and Leroy, J.-M.,** Hydrogenation of CO_2 over Co/Cu/K catalysts, *J. Chem. Soc., Far. Trans. I,* 83, 1711–1718, 1987.

29. **Thampi, K. R., Kiwi, J., and Grätzel, M.,** Methanation and photo-methanation of CO_2 at room temperature and atmospheric pressure, *Nature,* 327, 506–508, 1987.

30. **Grätzel, M.,** Catalytic and photocatalytic fixation of carbon dioxide, Proc. Int. Symp. Chemical Fixation of Carbon Dioxide, Nagoya, Japan, Dec. 2–4, 1991, 1–10.

31. **Melsheimer, J., Guo, W., Ziegler, D., Wesemann, M., and Schlögl, R.,** Methanation of carbon dioxide over Ru/Titania at room temperature — exploration for a photoassisted catalytic reaction, *Catal. Lett.,* 11, 157–168, 1991.

32. **Ogura, K., Kawano, M., and Adachi, D.,** Dark catalytic reduction of CO_2 over photo-pretreated NiO/kieselguhr catalyst, *J. Mol. Catal.,* 72, 173–179, 1992.

33. **Prairie, M. R., Renken, A., Highfield, J. G., Thampi, K. R., and Grätzel, M.,** A Fourier transform infrared spectroscopic study of CO_2 methanation on supported ruthenium, *J. Catal.,* 129, 130–144, 1991.

34. **Halmann, M.,** Comparative evaluation of catalytic, electrochemical, photosynthetic and biomimetic CO_2 fixation — an answer to the greenhouse effect?, Proc. Int. Symp. Chemical Fixation of Carbon Dioxide, Nagoya, Japan, Dec. 2–4, 1991, 129–138.

35. **Huang, S., Jiang, B.-N., and Yu, W.,** Kinetics of hydrogenation of CO_2 on supported metallic catalysts, Proc. Int. Symp. Chemical Fixation of Carbon Dioxide, Nagoya, Japan, Dec. 2–4, 1991, 145–150.

36. **Inui, T., Funabiki, M., and Takegami, Y.,** Simultaneous methanation of carbon monoxide and carbon dioxide on supported nickel-based composite catalysts, *Ind. Eng. Chem. Prod. Res. Dev.,* 19, 385–388, 1980.

37. **Inui, T.,** Effective conversion of CO_2 and H_2 to hydrocarbons, Proc. Int. Symp. Chemical Fixation of Carbon Dioxide, Nagoya, Japan, Dec. 2–4, 1991, 159–166.

38. **Dubois, J.-L., Sayama, K., and Arakawa, H.**, CO_2 hydrogenation over carbide catalysts, *Chem. Lett.*, 1992, 5–8.

39. **Schild, C., Wokaun, A., Koeppel, R. E., and Baiker, A.**, CO_2 hydrogenation over nickel/zirconia catalysts from amorphous precursors: On the mechanism of methane formation, *J. Phys. Chem.*, 95, 6341–6346, 1991.

40. **Reller, A., Padeste, C., and Hug, P.**, Formation of organic carbon compounds from metal carbonates, *Nature*, 329, 527–529, 1987.

41. **Tsuneto, A., Kudo, A., Saito, N., and Sakata, T.**, Hydrogenation of solid state carbonates, *Chem. Lett.*, 1992, 831–834.

42. **Wade, L. E., Gengelbach, R. B., Trumbley, J. L., and Hallbauer, W. L.**, in *Kirk-Othmer Encyclopedia of Chemical Technology*, Vol. 15, 3rd ed., Wiley-Interscience, New York, 1981, 398–415.

43. **Ziessel, R.**, Chimie de coordination de la molecule de dioxyde de carbon: activation biologique, chimique, electrochimique et photochimique, *Nouv. J. Chim.*, 7, 613–633, 1983.

44. **Bardet, R., Thivolle-Cazat, J., and Trambouze, Y.**, Methanolation of carbon oxides on Cu-Zn catalysts at atmospheric pressure, *J. Chim. Phys.*, 78, 135–138, 1981.

45. **Denise, B., Sneeden, R. P. A., and Hamon, C.**, Hydrocondensation of carbon dioxide. IV, *J. Mol. Catal.*, 17, 359–366, 1982.

46. **Baiker, A. and Gasser, D.**, Supported palladium catalyst prepared from amorphous palladium-zirconium, *J. Chem. Soc., Faraday Trans. 1*, 85, 999–1007, 1989.

47. **Schild, C., Wokaun, A., and Baiker, A.**, On the mechanism of CO and CO_2 hydrogenation reactions on zirconia-supported catalysts: a diffuse reflectance FTIR study. I. Identification of surface species and methanation reactions on palladium/zirconia catalysts, *J. Mol. Catal.*, 63, 223–242, 1990.

48. **Schild, C., Wokaun, A., and Baiker, A.**, On the mechanism of CO and CO_2 hydrogenation reactions on zirconia-supported catalysts: a diffuse reflectance FTIR study. II. Surface species on copper/zirconia catalysts: implications for methanol synthesis selectivity, *J. Mol. Catal.*, 63, 243–254, 1990.

49. **Schild, C., Wokaun, A., and Baiker, A.**, On the hydrogenation of CO and CO_2 over copper/zirconia and palladium/zirconia catalysts, *Fresenius J. Anal. Chem.*, 341, 395–401, 1991.

50. **Koeppel, R. A., Baiker, A., and Wokaun, A.**, Copper zirconia catalysts for the synthesis of methanol from carbon dioxide: influence of preparation variables on structural and catalytic properties of catalysts, *Appl. Catal. A*, 84, 77–102, 1992.

51. **Koeppel, R. A., Baiker, A., Schild, C., and Wokaun, A.**, Carbon dioxide hydrogenation over Au/ZrO_2 catalysts from amorphous precursurs — catalytic reaction mechanisms, *J. Chem. Soc., Faraday Trans.*, 87, 2821–2828, 1991.

52. **Iizuka, T., Kojima, M., and Tanabe, K.**, Support effects in the formation of methanol from CO_2 and H_2 over rhenium catalysts, *J. Chem. Soc., Chem. Commun.*, 1983, 638–639.

53. **Inoue, T., Iizuke, T., and Tanabe, K.**, Support effect of zinc oxide catalysts on synthesis of methanol from CO_2 and H_2, *Bull. Chem. Soc. Jpn.*, 60, 2663–2664, 1987.

54. **Klier, K., Chatikavanij, V., Herman, R. G., and Simmons, G. W.**, Catalytic synthesis of methanol from carbon monoxide/hydrogen. IV. The effect of carbon dioxide, *J. Catal.*, 74, 343–360, 1982.

55. **Klier, K.**, Catalytic conversions of CO_2: activation and hydrogenation, Proc. Int. Symp. Chemical Fixation of Carbon Dioxide, Nagoya, Japan, Dec. 2–4, 1991, 139–134.

56. **Chinchen, G. C., Denny, P. J., Parker, D. G., Spencer, M. S., and Whan, D. A.,** Mechanism of methanol synthesis from carbon dioxide/carbon monoxide/hydrogen mixtures over copper/zinc oxide/alumina catalysts: use of carbon-14-labeled reactants, *Appl. Catal.,* 30, 333–338, 1987.

57. **Rozovskii, A. Y., Lin, G. I., Liberov, L. G., Slivinskii, E. V., Loktev, S. M., Kagan, Y. B., and Bashkirov, A. N.,** Mechanism of methanol synthesis from carbon dioxide and hydrogen. III. Determination of the rates of individual steps using carbon-14 monoxide, *Kinet. Katal.,* 18, 691–699, 1977; *Chem. Abstr.,* 87, 117351c.

58. **Kuznetsov, V. D., Shub, F. S., and Temkin, M. J.,** Role of carbon dioxide in the synthesis of methanol in the presence of SNM-1 copper catalyst, *Kinet. Katal.,* 23, 932–935, 1982; *Chem. Abst.,* 97, 215266e.

59. **Saussey, J., Lavalley, J. C., and Bovet, C.,** Infrared study of carbon dioxide adsorption on zinc oxide. Adsorption sites, *J. Chem. Soc., Faraday Trans. 1,* 78, 1457–1463, 1982.

60. **Deluzarche, A., Hindermann, J. P., Kienemann, A., and Kieffer, R.,** Application of chemical trapping to the determination of surface species and to the study of their evolution under reaction conditions in heterogeneous catalysis, *J. Mol. Catal.,* 31, 225–250, 1985.

61. **Kieffer, R.,** Methanol synthesis from CO_2 + H_2, Proc. Int. Symp. Chemical Fixation of Carbon Dioxide, Nagoya, Japan, Dec. 2–4, 1991, 151–158.

62. **Chanchlani, K. G., Hudgins, R. R., and Silveston, P. L.,** Methanol synthesis from H_2, CO, and CO_2 over Cu/ZnO catalysts, *J. Catal.,* 136, 59–75, 1992.

63. **Fujita, S., Usui, M., Ohara, E., and Takezawa, N.,** Methanol synthesis from carbon dioxide at atmospheric pressure over Cu/ZnO catalyst: role of methoxide species formed on ZnO support, *Catal. Lett.,* 13, 349–358, 1992.

64. **Ghazi, M., Barrault, J., and Ménézo, J. C.,** CO_2 hydrogenation into methanol on supported nickel-molybdenum catalysts, *Rev. Trav. Chim. Pays-Bas,* 110, 19–22, 1991.

65. **Xu, Z., Qian, Z., Mao, L., Tanabe, K., and Hattori, H.,** Methanol synthesis from CO_2 and H_2 over CuO-ZnO catalysts combined with metal oxides under 13 atm pressure, *Bull. Chem. Soc. Jpn.,* 64, 1658–1663, 1991.

66. **Xu, Z., Qian, Z., Tanabe, K., and Hattori, H.,** Support effect of Re catalyst on methanol synthesis from CO_2 and H_2 under pressure of 5 atm, *Bull. Chem. Soc. Jpn.,* 64, 1664–1668, 1991.

67. **Xu, Z., Quian, Z. H., and Hattori, H.,** Mechanistic study of the hydrogenation of carbon dioxide over supported rhenium and copper-zinc catalysts, *Bull. Chem. Soc. Jpn.,* 64, 3432–3437, 1991.

68. **Inui, T., Funabiki, M., and Takegami, Y.,** Dynamic adsorption studies on nickel-based methanation catalysts using the continuous flow method, *J. Chem. Soc., Faraday Trans. 1,* 76, 2237–2250, 1980.

69. **Inui, T. and Takeguchi, T.,** Effective conversion of carbon dioxide and hydrogen to hydrocarbons, *Catalysis Today,* 10, 95–106, 1991.

70. **Fujiwara, M. and Souma, Y.,** Hydrocarbon synthesis from carbon dioxide and hydrogen over Cu-Zn-Cr oxide zeolite hybrid catalysts, *J. Chem. Soc., Chem. Commun.,* 1992, 767.

71. **Calverley, E. M. and Smith, K. J.,** The effects of carbon dioxide, methanol, and alkali promoter concentration on the higher alcohol synthesis over a Cu/ZnO/Cr_2O_3 catalyst, *J. Catal.,* 130, 616–626, 1991.

72. **Shustorovich, E. and Bell, A. T.,** An analysis of methanol synthesis from CO and CO_2 on Cu and Pd surfaces by the bond-order-conservation-Morse-potential approach, *Surface Sci.,* 253, 386–394, 1991.

73. **Arakawa, H., Dubois, J.-L., and Sayama, K.,** Selective conversion of CO_2 to methanol by catalytic hydrogenation over promoted copper catalyst, Proc. First Int. Conf. on Carbon Dioxide Removal, Amsterdam, March 4–6, 1992.

74. **Tagawa, T., Shimakage, M., and Goto, S.,** Copper-based supported catalysts for methanol synthesis from CO_2 + H_2, Proc. Int. Symp. Chemical Fixation of Carbon Dioxide, Nagoya, Japan, Dec. 2–4, 1991, 409–412.

75. **Tsai, J.-C. and Nicholas, K. M.,** Transition metal-mediated photochemical and catalytic reduction of carbon dioxide: rhodium-catalyzed reduction of carbon dioxide to formic acid, Proc. Int. Symp. Chemical Fixation of Carbon Dioxide, Nagoya, Japan, Dec. 2–4, 1991, 281–286.

76. **Tsai, J.-C. and Nicholas, K. M.,** Rodium-catalyzed hydrogenation of carbon dioxide to formic acid, *J. Am. Chem. Soc.,* 114, 5117–5124, 1992.

77. **Stalder, C. J., Chao, S., Summers, D. P., and Wrighton, M. S.,** Supported palladium catalysts for the reduction of sodium bicarbonate to sodium formate in aqueous solution at room temperature and one atmosphere of hydrogen, *J. Am. Chem. Soc.,* 105, 6318–6320, 1983.

78. **Wiener, H., Blum, J., Feilchenfeld, H., Sasson, Y., and Zalmanov, N.,** The heterogeneous catalytic hydrogenation of bicarbonate to formate in aqueous solutions, *J. Catal.,* 110, 184–190, 1988.

79. **Graf, E. and Leitner, W.,** Direct formation of formic acid from carbon dioxide and dihydrogen using the [{Rh(cod)Cl}$_2$]-Ph$_2$P(CH$_2$)$_4$PPh$_2$ catalyst system, *J. Chem. Soc., Chem. Commun.,* 1992, 623–624.

80. **Darensbourg, D. J., Ovelles, C., and Pala, M.,** Homogeneous catalysts for carbon dioxide/hydrogen activation. Alkyl formate production using anionic ruthenium carbonyl clusters as catalysts, *J. Am. Chem. Soc.,* 105, 5937–5939, 1983.

81. **Darensbourg, D. J., Pala, M., and Waller, J.,** Potential intermediates in carbon dioxide reduction processes. Synthesis and structure of (μ-formato) decacarbonyl triruthenium and (μ-acetato) decacarbonyl triruthenium anions, *Organometallics,* 2, 1285–1291, 1983.

82. **Taoda, H., Osaki, T., Horiuchi, T., Iseda, K., Tsuge, A., and Yamakita, H.,** Catalytic reduction of carbon dioxide by supported molybdenum sulphide, Proc. Int. Symp. Chemical Fixation of Carbon Dioxide, Nagoya, Japan, Dec. 2–4, 1991, 379–382.

83. **Ernst, K.-H., Campbell, C. T., and Moretti, G.,** Kinetics of the reverse water-gas shift reaction over Cu(110), *J. Catal.,* 134, 66–74, 1992.

84. **Yoshida, T., Thorn, D. L., Okano, T., Ibers, J. A., and Otsuka, S.,** Hydration and reduction of carbon dioxide by rhodium hydride compounds. Preparation and reactions of rhodium bicarbonate and formate complexes, and the molecular structure of RhH$_2$(O$_2$COH)(i-Pr)$_3$)$_2$, *J. Am. Chem. Soc.,* 101, 4212–4221, 1979.

85. **Tanaka, K., Okabe, J., and Aomura, K.,** A stoicheiometric conversion of CO + CH$_4$ into 2CO + 2H$_2$ by microwave discharge, *J. Chem. Soc., Chem. Commun.,* 1982, 921–922.

86. **Wan, J. K. S., Bamwenda, G., and Depew, M. C.,** Microwave induced catalytic reactions of carbon dioxide and water — mimicry of photosynthesis, *Res. Chem. Intermed.,* 16, 241–255, 1991.

87. **Henglein, A.,** Sonolysis of carbon dioxide, nitrous oxide and methane in aqueous solution, *Zeit, Naturforsch.,* 40B, 100–107, 1985.

88. **Yokoyama, S., Miyahara, K., Tanaka, K., Takakuwa, I., and Tashiro, J.,** Catalytic reduction of carbon dioxide. I. Reduction of carbon dioxide with carbon carrying potassium carbonate, *Fuel,* 58, 510–513, 1979.

89. **Carrasco-Marín, F., Rivera-Utrilla, J., Utrera-Hidalgo, E., and Moreno-Castilla, C.,** MoO_2 as catalyst in the CO_2 gasification of activated carbons and chars, *Fuel,* 70, 13–16, 1991.

90. **Dhupe, A. P., Gokarn, A. N., and Doraiswamy, L. K.,** Investigations into the compensation effect at catalytic gasification of active charcoal by carbon dioxide, *Fuel,* 70, 839–844, 1991.

91. **Yang, R. T. and Wong, C.,** Fundamental differences in the mechanism of carbon gasification by steam and by carbon dioxide, *J. Catal.,* 82, 245–251, 1983.

92. **Hüttinger, K. J. and Fritz, O. W.,** The carbon-carbon dioxide reaction: an extended treatment of the active site concept, *Carbon,* 29, 1113–1118, 1991.

SIX

Photochemical Reduction

The photochemical fixation of carbon dioxide has stimulated considerable interest, in providing models of prebiotic photosynthesis.[1] Earlier work on such artificial photosynthesis has been reviewed.[2]

UV ILLUMINATION IN HOMOGENEOUS MEDIA

The direct photolysis of CO_2 in the homogeneous gas phase occurs only by excitation in the vacuum UV region, at $\lambda = 120$ to 167 nm. The reaction steps involved have been represented by

$$CO_2 + h\nu \rightarrow CO + O(^1D) \qquad (1)$$

$$O(^1D) + CO_2 \rightarrow O(^3D) + CO_2 \qquad (2)$$

$$2O(^3D) + X \rightarrow O_2 + X \qquad (3)$$

where X is either the vessel wall, or some third body. The quantum yield of the overall reaction,

$$CO_2 \rightarrow CO + 1/2O_2 \qquad (4)$$

was found to be unity.[3]

The photolysis of gaseous mixtures of carbon dioxide and water was performed by vacuum UV irradiation ($\lambda = 147$ nm), leading to carbon monoxide and hydrogen. In the total gas pressure range of 4 to 29 kPa, at 0.1% H_2O in CO_2, the total quantum yield was $\Phi(CO) + \Phi(H_2) = 1$.[4]

Early studies on the reduction of aqueous solutions of carbon dioxide by UV light indicated the production of CO, formic acid, and formaldehyde.[5] The yield of the organic products was markedly enhanced in the presence of ferrous ions.[6]

For aqueous bicarbonate ions, the UV absorption onset occurs below about 240 nm, resulting in photoionization and formation of the bicarbonate radical, which is in acid-base equilibrium with the carbonate radical-anion,

$$HCO_3^- + h\nu \rightarrow HCO_3\cdot + e_{aq}^- \tag{5}$$

$$HCO_3\cdot + H_2O \leftrightarrow CO_3^-\cdot + H_3O^+ \quad pK_a = 7.9 \tag{6}$$

As noted in Chapter 3, the carbonate radical anion has been detected in flash photolysis by its transient absorption spectrum, with $\lambda_{max} = 600$ nm and $\epsilon = 191$ mol^{-1} m^2.[7,8]

In the presence of uranyl ions, steady state or flash photolytic illumination ($\lambda > 248$ nm) of carbonate or bicarbonate solutions resulted in photochromic behavior, which was postulated due to the formation of a radical complex intermediate,[9]

$$UO_2(CO_3)_2(CO_3\cdot)^{4-}$$

HOMOGENEOUS PHOTOSENSITIZED REACTIONS

Intensive efforts have been made during the last decade to mimic the natural process of photosynthesis, using reaction systems which included photosensitizers (such as porphyrin or bipyridine derivatives), electron transfer mediators (such as methyl viologen), and sometimes sacrificial electron donors (such as tertiary amines). The need for sacrificial electron donors is due to the fact that the reduction products of carbon dioxide, such as formic acid, formaldehyde, and methanol, are themselves very effective reducing agents.

Multielectron transfer in the photoreduction of carbon dioxide to methane was demonstrated using dititano-decatungstophosphate as electron transfer agent. With $[PTi_2W_{10}O_{40}]^{7-}$ (10 mM) in aqueous CO_2 saturated solutions, containing ethanol (2.5 M) as electron donor, by illumination ($\lambda > 270$ nm) of its intense UV absorption band ($\lambda_{max} = 255$ nm, $\epsilon = 3.9 \times 10^4$ M^{-1} cm^{-1}) resulted in the production of hydrogen, formaldehyde, and methane. Quantum yields for illumination at 313 nm were $\Phi_{HCHO} = 4 \times 10^{-3}$ and $\Phi_{CH4} = 6 \times 10^{-4}$.[10]

To Carbon Monoxide

The photoinduced reduction of carbon dioxide and water was observed by visible light illumination of Ru(2,2'-bipyridine)$_3^{2+}$, cobalt(II) chloride, and carbon dioxide in acetonitrile/water/triethylamine solution, yielding simultaneously carbon monoxide and water. In this system, the ruthenium complex was the photosensitizer, triethylamine served as sacrificial electron donor, and $CoCl_2$ acted as electron mediator. The amount of the product synthesis gas, and the CO/H_2 ratio, depended on the composition of the reactants. If triethanolamine was substituted for triethylamine, the selectivity for CO production was markedly enhanced.[11-13]

With (2,2'-bipyridine) tricarbonyl chlororhenium(I) as photosynthesizer in a similar homogeneous photochemical system, a quantum yield of 14% for the generation of CO was achieved.[14] With the same Re-(2.2'-bipyridine)(CO)$_3$Cl complex as photosensitizer, in dimethyl formamide containing tetraethyl ammonium chloride as medium, with triethanolamine as sacrificial electron donor, the photoreduction of CO_2 to CO occurred with a turnover number of 23 within 4 h. In a similar system, but with Re-p-(2,6-di-2-pyridyl-4-pyridyl) phenol as sensitizer, and dimethyl sulfoxide as solvent, the turnover number was 5.[15] Efficient carbon dioxide reduction to carbon monoxide in short-time photolysis was achieved using Ru(II)(bipy)$_3$Cl as a photosensitizer and Re(CO)$_3$(bipy)Cl as a cocatalyst. However, this system ceased to produce CO after 5 to 7 h photolysis, and instead released hydrogen.[16] The intermediates in CO_2 reduction catalyzed by rhenium tricarbonyl bipyridyl derivatives were identified in an *in situ* infrared study.[17]

Tetraaza-macrocyclic complexes are interesting for CO_2 fixation, since it had been observed that, in the natural reduction of CO_2 to CH_4 by methanogenic bacteria, one of the active coenzymes is a nickel tetrapyrrole.[18] Tetraaza-macrocyclic cobalt(II) complexes were used as electron transfer mediators in the photoreduction of carbon dioxide, with Ru(2,2'-bipyridine)$_3^{2+}$ as sensitizer, ascorbic acid as sacrificial electron donor, in aqueous CO_2 saturated solutions (pH 4). Such Co(II) complexes involve coordination of the macrocyclic ligands in a square planar configuration. The CO_2 group may possibly coordinate in the axial sites. Under illumination with daylight lamps, using the complex [Co(II)(Me$_2$(14)-4,11-dieneN$_4$]$^{2+}$.

the only observed products were CO and H_2, formed in the ratio 0.27:1. The turnover number (moles of CO + H_2 per mole of the complex) exceeded 500.[19] With a similar homogeneous photocatalyst system, using [Ru(bipy)$_3$]$^{2+}$ as photosensitizer and ascorbate buffer as the electron donor, but with $[Ni(cyclam)]^{2+}$ as the catalyst, enhanced CO/H_2 ratios were achieved. The CO and H_2 yields depended on the pH of the medium. Optimal CO production rate was at pH 5, providing a CO/H_2 ratio of 0.83. The mechanism was proposed to involve protonation of the reduced Ni species, followed by CO_2 insertion into the Ni-H bond, and finally dissociation to release CO and H_2O.[18] With the nickel cyclam $[Ni(14-aneN_4)]^{2+}$ as catalyst, $Ru(bipy)_3^{3+}$ as photosensitizer, and ascorbate as buffer (pH 5) and as sacrificial electron donor, at $\lambda = 400$ nm, the quantum yield for CO production was 0.06%. In addition to the problem of low quantum yield, the $Ru(bipy)_3^{2+}$ sensitizer was reported to undergo quite rapid decomposition, 25% within 4 h of photolysis. The primary chemical reaction following the photoexcitation of the sensitizer was proposed to be reductive quenching of the ruthenium excited state by the ascorbate anion.[20]

To Formic Acid

A system mimicking the electron transport sensitization in natural photosynthesis included an aromatic hydrocarbon such as pyrene or perylene, which transferred electrons from its singlet excited state to an electron acceptor such as 1,4-dicyanobenzene or 9,10-dicyanoanthracene in a polar medium, such as acetonitrile-water (5:1). The acceptor was then presumed to transfer an electron to carbon dioxide, forming the intermediate CO_2^- anion radical, which underwent protonation and further reduction to formate. Hydrogen peroxide was an additional product of the reaction.[21] However, in a reinvestigation of this reaction, it was observed that, if oxygen was carefully excluded from the reaction medium, no formic acid was produced. It was thus necessary to conclude that the origin of the formic acid was not carbon dioxide. Since the photosensitizer was consumed during the illuminations, it was proposed that the origin of the formic acid was the photooxidation of the aromatic sensitizers.[22]

The photoreduction of carbon dioxide to formic acid was achieved even in the absence of an electron mediator, using *p*-terphenyl as photocatalyst, in an aprotic polar solvent such as DMF, with triethylamine as a sacrificial electron donor — leading to a quantum yield of HCOOH production of 7.2% at 313 nm.[23] Considerable enhancement in the selectivity of *p*-terphenyl catalyzed photoreduction of CO_2 to CO was obtained by the addition of cobalt(III) cyclam (cyclam = 1,4,8,11-tetra-azacyclo tetradecane) as electron mediator. Using this cyclam, in acetonitrile-methanol solution, in the presence of triethanolamine as sacrificial electron donor, the apparent quantum yields at 313 nm for production of CO and HCOOH were 15 and 10%, respectively.[24,25]

In a comparison of oligo(*p*-phenylenes), OPP-n, of different chain lengths as catalysts for the photoreduction of CO_2 to formic acid and carbon monoxide, the most effective catalyst was the tetramer, OPP-4. These reactions were performed in nonaqueous solvents, preferably DMF, in the presence of triethylamine as a sacrificial electron donor, and without any electron transfer mediator, by illumination at $\lambda > 290$ nm. With OPP-4 as photocatalyst, the turnover number based on the consumed photocatalyst was 45, the apparent quantum yield at 313 nm was $\Phi_{HCOOH} = 0.084$. The reaction depended on photoexcitation of the oligo(*p*-phenylenes). For OPP-4, the absorption maximum $\lambda_{max} = 313$ nm.[26] The reduction potential of OPP-4 in dimethylamine solution had previously been determined by cyclic voltammetry to be -2.23V vs. SCE,[27] so that this oligomer in its reduced form should be able to transfer one electron to CO_2, since E° for the CO_2/CO_2^- couple is -2.21V in DMF solution.[28]

UV illumination at $\lambda = 254$ nm of CO_2 saturated aqueous solutions containing $Fe(II)(bpy)_3^{2+}$ (bpy = 2,2'-bipyridyl) resulted in the production of formic acid and formaldehyde. The yield of formaldehyde, based on the amount of Fe(II) that had been consumed, was about 0.06%.[29]

With the ruthenium complex $Ru(bpy)_3$ as photosensitizer, methylviologen as electron relay, and triethanolamine or EDTA as sacrificial electron donor, in carbon dioxide-saturated aqueous solution, formic acid production occurred with a quantum yield of about 1%.[30]

In the photochemical reduction of CO_2 by triethanolamine in DMF solution, using $[Ru(bpy)_3]^{2+}$ as photosensitizer, HCOOH was the major product. The active catalyst in this reaction was proposed to be the intermediate complex $[Ru(bpy)_2(CO)H]^+$.[31]

Carbon dioxide reduction to formate with a maximal quantum yield of 15% was obtained using a mixture of $[Ru(bpy)_3]^{2+}$ catalysts with triethanolamine as a sacrificial electron donor in DMF as solvent.[32]

An analogous photochemical system with $[Ru(bpy)_2(CO)_2]^{2+}$ as catalyst and triethanolamine in DMF as solvent and electron donor resulted in a maximal quantum yield of 14% for formate production.[33] A carbonyl carbon of this catalyst was shown to undergo nucleophilic attack by OH^- in neutral aqueous solutions,

$$[Ru(bpy)_2(CO)_2]^{2+} + OH = [Ru(bpy)_2(CO)C(O)OH]^+ \qquad (7)$$

$$[Ru(bpy)_2(CO)C(O)OH]^+ + OH^- = [Ru(bpy)_2(CO)(COO)] + H_2O \qquad (8)$$

The $[Ru(bpy)_2(CO)(COO)]$ species was proposed to be a key intermediate in the reduction of CO_2. A crystal structure determination of the hydrate $[Ru(bpy)_2(CO)(COO)].3H_2O$ indicated that the ruthenium atom was octahedrally coordinated by a CO, an η^1-CO_2 in a cis position, and four nitrogen atoms of 2,2'-bipyridine ligands.[34]

The main problem with photochemical reactions in homogeneous solutions seems to be the need for sacrificial electron donors — as well as low quantum yields, mainly because of side reactions.

HOMOGENEOUS PHOTOCARBOXYLATION

Carbon dioxide may be incorporated into aromatic compounds by illumination in the presence of a sensitizer and an electron donor. Solutions of phenanthrene and amines (such as dimethyl or diethyl aniline) and CO_2 (pressurized to 4 kg cm^{-2})

in dimethyl sulfoxide or dimethylformamide solution under UV illumination with a high-pressure Hg lamp yielded 9,10-dihydro-phenanthrene-9-carboxylic acid, with conversions (based on the phenanthrene consumed) of up to 55%.[35]

In dimethylformamide solution, using N,N-dimethylaniline as donor, naphthalene was carboxylated by carbon dioxide under illumination with visible light, in the presence of phenazine as a sensitizer. The selectivity to naphthoic acids was up to 67%, of which 90% was 1-naphthoic acid.[36,37]

A unique reaction is the visible light-induced fixation of carbon dioxide with zinc porphyrins in the presence of secondary amines or alcohols to produce zinc porphyrin carbamates or carbonates. N-Methyltetraphenyl-porphinato-zinc-ethyl in benzene solution in the presence of a secondary amine such as diethylamine reacted with carbon dioxide at room temperature to produce methyltetraphenyl-porphinato-zinc diethylcarbamate. If the amino was replaced by ethanol, the corresponding porphinato-zinc ethyl carbonate was formed. This reaction, which occurred slowly even in the dark, was considerably accelerated by illumination with visible light ($\lambda > 420$ nm).[38]

HIGH ENERGY RADIATION INDUCED CARBOXYLATION

Ionizing radiation of 40-MeV helium ions from a cyclotron[39,40] or of γ-rays from a ^{60}Co source[5,6] caused reduction of carbon dioxide to carbon monoxide, formic acid, formaldehyde, oxalic acid, and glyoxalic acid. Radiation-induced carboxylation with $^{14}CO_2$ has been used to convert amines into amino acids.[41] Thus, the carboxylation of methylamine caused the formation of glycine and proline, with G-values of about 0.5 for each product (G-value = number of changed molecules per 100 eV of absorbed energy). The carboxylation of ethylamine resulted mainly in alanine, β-alanine, while n-propylamine was converted mainly into α-aminobutyric acid. From iso-amylamine, the main product was leucine.[41]

The radiation-induced carboxylation of methanol with carbon dioxide was tested both in aqueous solutions and in neat methanol. In aqueous solutions containing methanol, the primary reactions were considered to be the reduction of CO_2 by e_{aq}^- and by H atoms to form the CO_2^- radical anion. Organic compounds such as methanol react very rapidly with $\cdot OH$ radicals to form organic radicals, which may either dimerize or react with the CO_2^- radical anions,

$$RH + \cdot OH \rightarrow R\cdot + H_2O \qquad (9)$$

$$R\cdot + CO_2^- \rightarrow RCOO^- \qquad (10)$$

$$2R\cdot \rightarrow RR \qquad (11)$$

The main radiolytic products of 1 mol dm^{-3} methanol in aqueous solutions under elevated CO_2 pressures from 1 to 13 atm were formic and glycolic acids and ethylene glycol. In pure methanol containing 1.7 mol dm^{-3} CO_2, the final products were formic acid, glycolic acid, ethylene glycol, and formaldehyde, with the G values 3.0, 2.3, 1.6, and 0.7, respectively.[42]

γ-Irradiation of aqueous suspensions of Mg, Ca, K, Tl, Cu, and Hg carbonates resulted in the formation of oxalic acid as the major product, as well as traces of HCHO, HCOOH, and glycolic acid.[43] In a study of the effects of radiation on aqueous suspensions of semiconductors, the radiolysis of the CO_2-H_2O-TiO_2 system was investigated. The major products were oxalic acid and formaldehyde, as well as traces of formic acid.[44]

In the radiolysis of CO_2 containing 10% H_2, the radiation chemical yield of CO was independent of temperature up to 400°C. Raising the temperature from 500 to 700°C caused a sharp increase in G(CO), as well as consumption of H_2. At 700°C, full conversion of H_2 was observed. A chain reaction was postulated, with a chain length of 10 to 150 in the temperature range of 500 to 700°C.[45]

In the radiolysis of CO_2 over a ZnO catalyst, the rate of CO formation was faster than in the homogeneous gas phase decomposition. The initial radiation chemical yield of CO production was G = 4 mol/100eV. In the irradiated ZnO, paramagnetic Zn^+ and O^- centers were observed by ESR.[46]

The γ-radiolysis of CO_2-saturated aqueous solutions containing sodium phosphomolybdate, $Na_3H_4[P(Mo_2O_7)_6]$, resulted in the production of formaldehyde and oxalic acid.[47]

The radiolysis reactions of low levels of hydrocarbons were tested in mixtures of CO_2-CO-CH_4-H_2O-H_2, which are typical of conditions in a gas-cooled nuclear reactor. Reaction intermediates included clustered positive and negative ions, as well as O and H atoms and OH and alkyl radicals.[48]

References

1. **Chittenden, G. J. F. and Schwartz, A. W.,** Prebiotic photosynthetic reactions, *BioSystems,* 14, 15–32, 1981.

2. **Halmann, M.,** Photochemical fixation of carbon dioxide, in *Energy Resources Through Photochemistry and Catalysis,* Grätzel, M., Ed., Academic Press, New York, 1983, 507–534.

3. **Slanger, T. G. and Black, G.,** CO_2 photolysis revisited, *J. Chem. Phys.,* 68, 1844–1849, 1978.

4. **Kurbanov, M. A., Rustamov, V. R., Mustafaev, I. I., Iskanderova, Z. I., and Gadzhiev, Kh. M.,** Principles of the photolysis of gaseous mixtures of carbon dioxide and water, *Khim. Vys. Energ.,* 18, 381–382, 1984; *Chem. Abstr.,* 101, 101101t.

5. **Getoff, N., Scholes, G., and Weiss, J.,** Reduction of carbon dioxide in aqueous solutions under the influence of radiation, *Tetrahedron Lett.,* 1960, 17–23.

6. **Getoff, N.,** Reduction of carbonic acid in aqueous solution under the influence of UV-light, *Z. Naturforsch.,* 17b, 87–90, 1962.

7. **Eriksen, T. E., Lind, J., and Merenyi, G.,** On the acid-base equilibrium of the carbonate radical, *Radiat. Phys. Chem.,* 26, 197–199, 1985.

8. **Neta, P., Huie, R. E., and Ross, A. B.,** Rate constants for reactions of inorganic radicals in aqueous solutions, *J. Phys. Chem. Ref. Data,* 17, 1027–1284, 1988.

9. **Saini, R. D. and Iyer, R. M.,** On the reaction of carbonate radical with uranyl ion in aqueous medium: flash photolytic and pulse radiolytic studies, *J. Photochem. Photobiol. A,* 61, 171–182, 1991.

10. **Yamase, T. and Sugeta, M.**, Photoreduction of CO_2 to CH_4 in water using dititano-decatungstophosphate as multielectron transfer catalyst, *Inorg. Chim. Acta*, 172, 131–134, 1990.

11. **Lehn, J.-M. and Ziessel, R.**, Photochemical generation of carbon monoxide and hydrogen by reduction of carbon dioxide and water under visible light irradiation, *Proc. Natl. Acad. Sci. U.S.A.*, 79, 701–704, 1982.

12. **Hawecker, J., Lehn, J.-M., and Ziessel, R.**, Efficient photochemical reduction of CO_2 to CO by visible light irradiation of systems containing $Re(bpy)(CO)_3X$ or $Ru(bipy)_3^{2+}$-Co^{2+} combination as homogeneous catalysts, *J. Chem. Soc. Chem. Commun.*, 1983, 536–538.

13. **Ziessel, R., Hawecker, J., and Lehn, J.-M.**, Photogeneration of carbon monoxide and of hydrogen via simultaneous photochemical reduction of carbon dioxide by visible-light irradiation of organic solutions containing tris (2,2'-bipyridine) ruthenium(II) and cobalt(II) species as homogeneous catalysts, *Helv. Chim. Acta*, 69, 1065–1084, 1986.

14. **Hawecker, J., Lehn, J.-M., and Ziessel, R.**, Photochemical and electrochemical reduction of CO_2 to CO mediated by (2,2'-bipyridine) tricarbonyl chloro rhenium (I) and related complexes as homogeneous catalysts, *Helv. Chim. Acta*, 69, 1990–2012, 1986.

15. **Calzaferri, G., Hädener, K., and Li, J.**, Photoreduction and electroreduction of carbon dioxide by a novel rhenium(I) p-phenyl-terpyridine complex, *J. Photochem. Photobiol., A*, 64, 259–262, 1992.

16. **Hukkanen, H. and Pakkanen, T. T.**, Photochemical catalytic reduction of carbon dioxide by visible light using $Ru^{II}(bipy)_3$ and $Re(CO)_3(bipy)Cl$ as photocatalysts, *Inorg. Chim. Acta*, 114, 43–45, 1986.

17. **Christensen, P., Hamnet, A., Muir, A. V. G., and Timney, J. A.**, An *in situ* infrared study of CO_2 reduction catalyzed by rhenium tricarbonyl bipyridyl derivatives, *J. Chem. Soc. Dalton Trans.*, 1992, 1455.

18. **Grant, J. L., Goswami, K., Spreer, L. O., Otvos, J. W., and Calvin, M.**, Photochemical reduction of CO_2 to CO in water using a Ni(II) tetra-aza macrocycle complex as catalyst, *J. Chem. Soc. Dalton Trans.*, 1987, 2105–2109.

19. **Tinnemans, A. H. A., Koster, T. P. M., Thewissen D. H. M. W., and Mackor, A.**, Tetraaza-macrocyclic cobalt(II) and nickel(II) complexes as electron-transfer agents in the photo (electro) chemical and electrochemical reduction of carbon dioxide, *Rec. Trav. Chim. Pays-Bas*, 103, 288–295, 1984.

20. **Craig, C. A., Spreer, L. O., Otvos, J. W., and Calvin, M.**, Photochemical reduction of carbon dioxide using nickel tetraazamacrocycles, *J. Phys. Chem.*, 94, 7957–7960, 1990.

21. **Tazuke, S. and Kitamura, N.**, Photofixation of carbon dioxide to formic acid using water as hydrogen source, *Nature*, 275, 301–302, 1978.

22. **Legros, B. and Soumillion, J. Ph.**, Is carbon dioxide photoreducible by monoelectronic transfers under visible light?, *Tetrahedr. Lett.*, 26, 4599–4600, 1985.

23. **Matsuoka, S., Kohzuki, T., Pac, C. J., and Yanagida, S.**, Photochemical reduction of carbon dioxide to formate catalyzed by p-terphenyl in aprotic polar solvent, *Chem. Lett.*, 1990, 2047–2048.

24. **Matsuoka, S., Yamamoto, K., Pac, C., and Yanagida, S.**, Enhanced para-terphenyl-catalyzed photoreduction of CO_2 to CO through the mediation of Co(III)-cyclam complex, *Chem. Lett.*, 1991, 2099–2100.

25. **Yanagida, S. and Matsuoka, S.**, Efficient photochemical reduction of CO_2 catalyzed by oligo-(p-phenylenes), Proc. Int. Symp. Chemical Fixation of Carbon Dioxide, Nagoya, Japan, Dec. 2–4, 1991, 19–22.

26. **Matsuoka, S., Kohzuki, T., Pac, C. J., Yanagida, S., Takamuku, S., Kusaba, M., Nakashisha, N., and Yanagida, S.**, Photocatalysis of oligo(*para*-phenylenes). Photochemical reduction of carbon dioxide with triethylamine, *J. Phys. Chem.*, 96, 4437–4442, 1992.

27. **Meerholtz, K. and Heinze, J.**, Multiple reversible electrochemical reduction of aromatic hydrocarbons in liquid alkylamines, *J. Am. Chem. Soc.*, 111, 2325–2326, 1989.

28. **Lamy, E., Nadjo, L., and Savéant, J. M.**, Standard potential and kinetic parameters of the electrochemical reduction of carbon dioxide in dimethylformamide, *J. Electroanal. Chem.*, 78, 403–407, 1977.

29. **Åkermark, B., Eklund-Westlin, U., Baeckström, P., and Löf, R.**, Photochemical, metal-promoted reduction of carbon dioxide and formaldehyde in aqueous solution, *Acta Chem. Scand. B*, 34, 27–30, 1980.

30. **Kitamura, N. and Tazuke, S.**, Photoreduction of carbon dioxide to formic acid mediated by a methylviologen electron relay, *Chem. Lett.*, 1983, 1109–1112.

31. **Hawecker, J., Lehn, J. M., and Ziessel, R.**, Photochemical reduction of CO_2 to formate mediated by ruthenium bipyridine complexes as homogeneous catalysts, *J. Chem., Soc. Chem. Commun.*, 1985, 56–58.

32. **Lehn, J.-M. and Ziessel, R.**, Photochemical reduction of CO_2 to formate catalyzed by 2,2'-bipyridine — or 1,10-phenanthroline — ruthenium complexes, *J. Organometal. Chem.*, 382, 157–173, 1990.

33. **Ishida, I., Terada, T., Tanaka, K., and Tanaka, T.**, Photochemical CO_2 reduction catalyzed by $[Ru(bpy)_2(CO)_2]_2^+$ using triethanolamine and 1-benzyl-1,4-dihydronicotinamide as an electron donor, *Inorg. Chem.*, 29, 905–911, 1990.

34. **Tanaka, H., Nagao, H., Peng, S.-M., and Tanaka, K.**, Crystal structure of cis-(carbon monoxide) (η^1-carbon dioxide) bis (2,2'-bipyridyl) ruthenium, an active species in catalytic CO_2 reduction affording CO and $HCOO^-$, *Organometallics*, 11, 1450–1451, 1992.

35. **Tazuke, S. and Ozawa, H.**, Photofixation of carbon dioxide: formation of 9,10-dihydrophenanthrene-9-carboxylic acid from phenanthrene-amine-carbon dioxide systems, *J. Chem. Soc. Chem. Commun.*, 1975, 237–238.

36. **Tagaya, H., Onuki, M., Tomioka, Y., Wada, Y., Karasu, M., and Chiba, K.**, Photocarboxylation of naphthalene in the presence of carbon dioxide and an electron donor, *Bull. Chem. Soc. Jpn.*, 63, 3233–3237, 1990.

37. **Tagaya, H., Onuki, M., Karasu, M., and Chiba, K.**, Photocarboxylation of an aromatic compound in the presence of carbon dioxide and an electron donor, Proc. Int. Symp. Chemical Fixation of Carbon Dioxide, Nagoya, Japan, Dec. 2–4, 1991, 195–200.

38. **Inoue, S., Nukui, M., and Kojima, F.**, Light-induced fixation of carbon dioxide with zinc porphyrin, *Chem. Lett.*, 1984, 619–622.

39. **Garrison, W. M., Morrison, D. C., Hamilton, J. G., Benson, A. A., and Calvin, M.**, Reduction of carbon dioxide in aqueous solutions by ionizing radiation, *Science*, 114, 416–418, 1951.

40. **Garrison, W. M. and Rollefson, G. K.**, Radiation chemistry of aqueous solutions containing both ferrous ions and carbon dioxide, *Discuss. Faraday. Soc.*, 1952, 155–161.

41. **Getoff, N., Gütlbauer, F., and De la Paz, L. R.**, Radiation induced preparation of labelled compounds. III. Incorporation of $^{14}CO_2$ and of $H^{14}COO^-$ into amines, *Kerntechnik*, 14, 75–81, 1972.

42. **Fjodorov, V. V. and Getoff, N.**, Radiation induced carboxylation of methanol under elevated CO_2-pressure, *Radiat. Phys. Chem.*, 22, 841–848, 1983.

43. **Lysyak, T. V., Konash, E. A., Kalyazin, E. P., Rudnev, A. V., and Kharitonov, Yu. Ya.,** Formation of organic products from metal carbonates and water in the presence of ionizing radiation, *Dokl. Akad. Nauk SSSR,* 265, 912–913, 1982; *Chem. Abstr.,* 97, 205662s.

44. **Kolomnikov, I. S., Lysyak, T. V., Konash, E. A., Rudnev, A. V., Kalyazin, E. A., and Kharitonov, Yu. Ya.,** Reduction of CO_2 in aqueous solution in the presence of TiO_2 under gamma radiation, *Zhur. Neorg. Khim. SSSR,* 28, 528–529, 1983; *Chem. Abstr.,* 98, 135105g.

45. **Kurbanov, M. A., Rustamov, V. R., Mamedov, Kh. F., Iskenderova, Z. I., and Dzantiev, B. G.,** Chain transformations during radiolysis of gaseous carbon dioxide-hydrogen mixtures, *Khim. Fiz.,* 5, 135–136, 1986; *Chem. Abstr.,* 104, 99290v.

46. **Rustamov, V. R., Kerimov, V. K., Kurbanov, M. A., and Ali-Zade, Sh. N.,** Heterogeneous radiolysis of carbon dioxide on a zinc oxide catalyst, *Khim. Vysok. Energii,* 19, 350–352, 1985; *Chem. Abstr.,* 103, 96184g.

47. **Lysyak, T. V., Konash, E. A., Rudnev, A. V., Kalyazin, E. P., Kolomnikov, I. S., and Kharitonov, Yu. Ya.,** Radiolysis of the CO_2-H_2O system in the presence of sodium phosphomolybdate, *Zh. Neorg. Khim. SSSR,* 28, 1603–1604, 1983; *Chem. Abstr.,* 99, 45964c.

48. **Norfolk, D. J., Skinner, R. F., and Williams, W. J.,** Hydrcarbon chemistry in irradiated $CO_2/CO/CH_4/H_2O/H_2$ mixtures. I. A survey of the initial reactions, *Rad. Phys. Chem.,* 21, 307–319, 1983.

SEVEN

Electrochemical Reduction

The equilibrium redox potentials $E°$ (vs. NHE) for the multielectron transfer reactions of carbon dioxide at pH 7.0 to formic acid, carbon monoxide, formaldehyde, methanol, and methane were calculated from the half-cell reactions.[1]

$$CO_2 + 2H^+ + 2e^- \rightarrow HCOOH \qquad (E° = -0.61V) \qquad \textbf{(1)}$$

$$CO_2 + 2H^+ + 2e^- \rightarrow CO + H_2O \qquad (E° = -0.52V) \qquad \textbf{(2)}$$

$$CO_2 + 4H^+ + 4e^- \rightarrow HCHO + H_2O \ (E° = -0.48V) \qquad \textbf{(3)}$$

$$CO_2 + 6H^+ + 6e^- \rightarrow CH_3OH + H_2O \ (E° = -0.38V) \qquad \textbf{(4)}$$

$$CO_2 + 8H^+ + 8e^- \rightarrow CH_4 + 2H_2O \quad (E° = -0.24V) \qquad \textbf{(5)}$$

A plot of redox potentials vs. the number of electrons transferred in the above reactions is presented in Figure 1, indicating that multielectron reductions are energetically favored.[1] Since these reactions require very much less energy per electron transferred than the direct monoelectronic reduction of carbon dioxide to its radical anion CO_2^-, i.e., -2.1 V vs. SCE,[2] there exists considerable interest to apply multielectron transfer to the reduction of carbon dioxide.

REDUCTION IN AQUEOUS SOLUTIONS

One of the first applications of electrochemical reactions to chemical synthesis has been the reduction of aqueous sodium bicarbonate to HCOOH.[3,4] Since then, the electrochemical reduction of carbon dioxide has been the subject of many

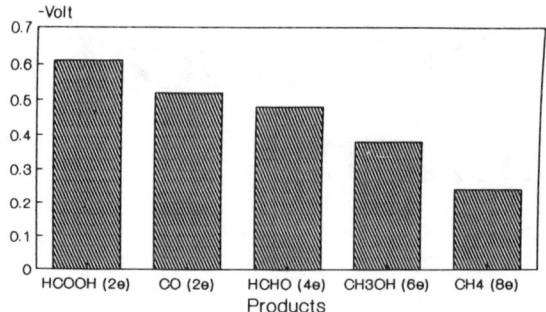

Fig. 1   Redox potentials vs. number of electrons transferred.

studies, in respect of the effect of the electrode materials and electrolyte media on the nature of the products, which include C_1 and sometimes C_2 and higher products. In particular, the nature of the electrochemical interaction of carbon dioxide with metals has been investigated by several physical techniques. The reaction was developed to a preparative method, using zinc amalgam electrodes, resulting in 89% Faradaic yield of formic acid.[5] Further enhancement in the Faradaic yield, to 95%, was achieved with ammonium carbonate as the electrolyte, and with the zinc amalgam electrode separated from the platinum anode by a clay diaphragm.[6] With similar zinc or lead amalgam electrodes, in saturated ammonium or potassium sulfate, but with CO_2 pressurized up to 50 atm in a high-pressure electrolysis cell, the conversion of CO_2 to formic acid was practically quantitative, and traces of methanol were also observed.[7] On sodium amalgam cathodes with 2% aqueous $NaHCO_3$ saturated with CO_2, at a current density of 16 to 25 mA cm^{-2}, the current yield of formic acid was 86 to 97%.[8] In a useful preparative procedure, rotating amalgamated copper cathodes in 10% sodium sulfate saturated with CO_2 were separated by a cation-exchange membrane from the lead electrode in the anodic compartment. At a cell potential of 3.5 V, the current density was 20 mA cm^{-2}, the Faradaic efficiency was 81%, and the concentration of formic acid reached up to 200 g l^{-1}.[9]

Reduction Mechanisms

On platinum metal in 2 N H_2SO_4, carbon dioxide was found by current-potential measurements to be adsorbed in a reduced form. In the potential range 0 to 250 mV (vs. NHE), CO_2 reacted with chemisorbed hydrogen on bright platinum to form a chemisorption product of reduced CO_2, which possibly may be CO. The rate of reduction of CO_2 increased markedly with the temperature. The oxidation of this reduction product was irreversible.[10] The nature of this reduced CO_2 on platinum was further investigated by looking for the adsorbed species using an *in situ* technique: electrochemically modulated infrared spectroscopy (EMIRS). The spectral data seemed to show that CO_2 was reduced by reaction with adsorbed hydrogen. Under potential modulation of the Pt electrode between $+250$ mV and $+550$ mV, two IR absorption bands were observed. The very strong band centered at about 2060 cm^{-1} was attributed to a linearly bonded CO species, while the moderately strong band at about 1865 cm^{-1} was assigned to a CO species adsorbed at a surface site of higher coordination number.[11] The above conclusions were confirmed in an FTIR spectroscopic study.[12]

The reduction of CO_2 to CO in aqueous 0.1 M HClO$_4$ on platinum single crystal electrodes was studied by voltammetry and Fourier transform infrared reflection absorption spectroscopy (FTIRRAS). On platinum electrodes the reduction of CO_2 occurred only in the hydrogen adsorption region. The data indicated that, on both the Pt(100) and Pt(110) surfaces, CO was the CO_2 reduction product, according to

$$CO_2 + H^+ + 2e^+ \rightarrow CO + H_2O \qquad (6)$$

The CO was adsorbed in two bonding states. On Pt(100), both linear CO and a large amount of bridge-bonded CO were observed, while on Pt(110) mainly linear CO and only a small amount of bridge-bonded CO were detected. The Pt(111) face exhibited only low activity for CO_2 reduction.[13] The effect of Cl$^-$ impurity on CO_2 reduction on polycrystalline Pt electrodes in 0.1 M HClO$_4$ was determined. Chloride ions are specifically adsorbed on Pt surfaces, and effect a modification in the hydrogen adsorption region — thus interfering in CO_2 reduction.[14]

On mercury cathodes in aqueous media, the reduction of carbon dioxide led to a polarographic wave with a half-wave potential of 2.1 V vs. the saturated calomel electrode (SCE).[2] On such electrodes, the primary step of carbon dioxide reduction to formate has been proposed to be electron capture by CO_2 to form the $\cdot CO_2^-$ ion radical in solution,

$$CO_2 + e_{aq}^- \rightarrow \cdot CO_2^- \ (E^\circ = -2.21 \text{ V vs. SCE}) \qquad (7)$$

An evaluation of the reduction potential of the CO_2/CO_2^- couple was made, using a thermodynamic cycle, and taking for the electron affinity of CO_2 EA $= -0.6$ eV. Thus, the value derived for E° $(CO_2/CO_2^-) = -1.93$ V vs. NHE.[15]

Using mercury[16] or tin and indium electrodes,[17] the rate constant for the reaction of the hydrated electron with the CO_2 molecule in aqueous solution was determined by kinetic measurements to be almost diffusion limited, with $k_2 = 7 \times 10^9$ M^{-1} s^{-1}.

The CO_2^- ion radical was identified by its strong optical absorption peak at 250 nm, using Pb electrodes in 0.1 M Me$_4$NCl saturated with CO_2, at a potential of -1.0 to -1.8 V (vs. Ag/AgCl), and applying modulated specular reflectance spectroscopy.[18] Photoemission studies on polarized mercury electrodes in contact with aqueous electrolytes indicated that the subsequent steps leading to formate ions occur on the electrode surface,[16]

$$H_2O + \cdot CO_2^- \rightarrow HCOO\cdot_{ads} + OH^- \quad k = 7.7 \times 10^2 \text{ M}^{-1} \text{ s}^{-1} \qquad (8)$$

$$HCOO\cdot_{ads} + e^- \rightarrow HCOO^- \qquad (9)$$

The current-voltage curves on Hg revealed two Tafel regions, with slopes of 0.09 V/decade and 0.22 V/decade, at low and high current densities. The changeover occurred at a current density of about 50 μA cm^{-2}. Reactions (9) and (7) have been proposed as the rate-determining steps in the low and high current density regions, respectively.[19-21]

Further studies on the reduction of carbon dioxide on mercury electrodes, using nanosecond pulsed laser or continuous wave (cw) photoelectron emission, showed that the emitted electrons were captured by CO_2, forming CO_2^-, which was adsorbed on the Hg surface. The current density of CO_2 reduction on Hg in neutral-buffered solutions was about 1 mA cm^{-2} at -1.6 V vs. SCE, with a reduction rate constant of $(3 \text{ to } 5) \times 10^6$ s^{-1}. The transfer coefficient was 0.25 to 0.3 in the potential range of -1.5 to -1.8 V.[22]

The kinetics and mechanism of CO_2 reduction to formic acid in aqueous solutions was studied on a wide range of metal electrodes with high and moderate hydrogen overvoltage, such as Sn, In, Bi, Sb, Cd, Zn, Cu, Pb, Ga, Ag, Au, Ni, Fe, W, Mo, as well as on glassy carbon. Optimal electrodes were those which exhibited CO_2 electroreduction near the zero-charge potential, in the range of maximal CO_2 adsorption. The electrode material strongly affected both of the two Tafel regions of the polarization curves. This suggested that both the first and the second electron were transferred to adsorbed species,

$$CO_2(ads) + e^- = CO_2^-(ads) \qquad (10)$$

$$CO_2^-(ads) + BH + e^- \rightarrow HCOO^- + B^- \qquad (11)$$

where BH was a proton donor, such as H_2O or H_3O^+. The surface coverage by adsorbed particles was very low on all the electrodes studied, $\phi < 0.1$.[23]

The adsorption of carbon dioxide on platinum group metals was measured on smooth Pd, Rh, Ir, Ru, and Os electrodes, in 1 N H_2SO_4, using the method of potentiodynamic pulses, with gold foil as the counterelectrode. On Ir, Pd, Ru, and Os electrodes, no chemisorption of CO_2 was observed. On Rh electrodes, the chemisorption of carbon dioxide was detected in the potential region of hydrogen adsorption. The activation energy for CO_2 adsorption on rhodium was 4.55 kcal. The species adsorbed on the rhodium electrode had the average composition of CH_2, and was so strongly bonded to the electrode surface that it was not desorbed during cathodic polarization.[24] The electroreduction of CO_2 on the platinum group metals Pt and Rh occurred by chemisorbed CO_2 species reacting with chemisorbed hydrogen on the metal surface.[25]

On platinum electrodes in dilute aqueous solutions carbon dioxide was not reduced to organic compounds.[26] However, in saturated carbonate electrolytes below $-5°C$ in one-compartment cells, there occurred appreciable production of formate ion and formaldehyde. These products were not formed in cells with separated compartments. The explanation given was that the percarbonate ion $C_2O_6^{2-}$ was produced on the anode, and was then reduced by hydrogenation on the cathode.[27]

On gold cathodes, the reduction of carbon dioxide led to CO as the main product, with Faradaic efficiency of about 60%, as well as hydrogen. A detailed kinetic investigation of the mechanism of CO_2 reduction at the Au electrode in phosphate buffer solutions (pH 2.5 to 6.8) suggested that the primary step, which is rate determining, is an electron transfer from the electrode to a surface-adsorbed CO_2 molecule,[28]

$$CO_2(ads) + e^- \rightarrow CO_2^-(ads) \qquad (12)$$

followed by rapid proton transfer from water, releasing carbon monoxide

$$CO_2^-(ads) + H_2O + e^- \rightarrow CO + 2OH^- \qquad (13)$$

On tin electrodes, the kinetics of the cathodic reduction of carbon dioxide was studied in more detail using both steady-state polarization curves recorded with a rotating disk electrode and photoelectron emission measurements at the metal-solution boundary. In contrast to the effect with Hg electrodes, the rate-determining step in reduction of CO_2 to formate on Sn cathodes in the first Tafel region was shown to be the transfer of the second electron. The mechanism was thus represented by[29]

$$CO_2 + e^- \rightarrow \cdot CO_{2\,ads}^- \qquad (14)$$

The high overvoltage encountered with mercury electrodes stimulated the search for other electrode materials. On tin cathodes, optimal conditions for formate production (92% current efficiency) were reported using a catholyte of 2% $KHCO_3$ continuously saturated with CO_2 (pH below 8.0) at a current density of 3.3 mA cm^{-1}.[30] Zinc, lead, tin, indium, and cadmium were studied as electrocatalysts for CO_2 reduction. Indium electrodes in 0.1 N lithium carbonate provided the largest current efficiency (92%) for the production of formic acid.[31]

The nature of the electroreduction products of carbon dioxide on zinc electrodes was found to depend on the electrolyte medium. In 0.1 M $KHCO_3$, at potentials of -1.5 to -1.7 V (vs. Ag/AgCl), both CO and HCOOH were formed, while in 0.05 M K_2SO_4 the predominant product was CO, in up to 80% Faradaic efficiency, with small yields of HCOOH. The explanation for this effect was the higher rate of dissolution of the Zn electrode in the K_2SO_4 solution (pH 4.2) than in the $KHCO_3$ medium (pH 6.8). The dissolved Zn^{2+} ions were found to promote the formation of carbon monoxide.[32]

Carbon dioxide reduction in aqueous solutions is enhanced by elevated CO_2 gas pressure, which provides increased solubility. The effect of carbon dioxide gas pressure on the current density was investigated in several studies. Lead, indium, zinc, and tin were tested as electrodes for the high gas pressure CO_2 reduction. In experiments with lithium carbonate as electrolyte, the current density increased with the gas pressure. However, the Faradaic efficiency rose rapidly with the pressure only up to 5 kg cm^{-2}, and then reached a plateau, while the current densities reached 30 mA cm^{-2}. Tin cathodes required the lowest overpotential, providing maximum current efficiency at a potential of -1.7 V (vs. Ag/AgCl).[33] With the metal electrodes Fe, Co, Ni, Rh, Pt, Mo, and Re, on which there was hardly any reduction of CO_2 at 1-atm pressure, significant reduction was observed at 50 atm of CO_2. Thus, on a nickel electrode in 0.1 M KHCO$_3$ solution, at an electrode potential of -1.8 V (vs. Ag/AgCl), at 1 and 50 atm of CO_2, the Faradaic efficiencies were for HCOOH, 0.1 and 12.0%, for CH$_4$, 0.6 and 1.2%, for C_2H_4, 0.6 and 0.03%, for C_2H_6, 0 and 0.7%, and for C_3H_8, 0.06 and 0.42%.[34,35]

In the reduction of carbon dioxide in neutral aqueous solutions (0.05 M phosphate buffer, pH 6.8) over a Hg-pool electrode, the Faradaic efficiency for formate production reached 81.5% at a cell potential of 3.5 V and a current density of 20 mA cm^{-2}. However, the energy efficiency (defined as the ratio of the free energy of the organic fuel produced to the electric energy input) was at best less than 45%. In the low overvoltage region, with a Tafel slope of about 91 mV, the galvanostatic charging curve indicated large coverage by an absorbed intermediate, presumably the above formate radical HCOO·$_{ads}$. On the other hand, this coverage was small in the high overvoltage region, with the Tafel slope of about 240 mV. The desired further reduction of formic acid to methanol was not possible directly in the above neutral medium. This was achieved in an acid medium, optimally at pH 3.8 (in 0.1 N NaHCO$_3$), on an electroetched Sn electrode, resulting in a Faradaic efficiency of up to 99.6%, but at a current density of only 3.6 μA cm^{-2}.[36]

On palladium electrodes, in contrast to mercury electrodes, the cathodic reduction was shown to occur directly on the bicarbonate ion, HCO$_3^-$, and not on CO_2. The reduction current was dependent on the cation of the electrolyte, decreasing in the order Cs > K > Na > Li. Thus, the current density in 1 M CsHCO$_3$ was nine times higher than in 1 M NaHCO$_3$. The cesium effect in promoting the reduction of HCO$_3^-$ was explained by competitive adsorption, decreasing the amount of adsorbed species from formaldehyde. Formaldehyde, which was presumably the reduction product of formate, was shown to cause the gradual blocking of the reduction of bicarbonate. An alternative interpretation was that Cs$^+$–HCO$_3^-$ ion pairs were directly involved in the reaction at the cathode.[37-39]

The electrochemical reduction of bicarbonate to formate was achieved with high current efficiency using Pd-impregnated polymers. The polymer was obtained from the monomer,

[MeO)3Si(CH2)3–N⟨=⟩–⟨=⟩N–(CH2)3Si(OMe)3]$^{2+}$

on metals (W or Pt). Highest current efficiency (85%) for the production of formate was obtained using 7 M $CsHCO_3$ as the catholyte, at a current density of about 0.1 mA cm^{-2}, and at a potential which was close to the thermodynamic potential E° (CO_3H^-/HCO_2^- = -0.76 V vs. SCE.[40]

In order to optimize the electroreduction of bicarbonate to formate on the redox polymers, it was necessary to determine the relative electrostatic binding of bicarbonate and formate ions on such polymers. The binding constants were measured by FTIR (Fourier transform infrared) spectroscopy of the polymer derived from an N,N'-dialkyl-4,4'-dipyridinium monomer immobilized on a single-crystal Si electrode surface. When both ions were at high concentrations, the CO_3H^- ion was more strongly bound. Thus, at 3.0 M concentration, the bicarbonate ion was about seven times more firmly bound that the formate ion.[41]

In a detailed study of the reduction products on a Pd wire electrode in CO_2-saturated aqueous solutions of 0.05 M $KHCO_3$, at -2.0 V vs. SCE, the Faradaic yields were for H_2 73%, HCOOH 16.1%, CO 11.6%, and CH_4 0.083%. Also, many saturated and unsaturated hydrocarbons, from C_1 to C_6, were detected in very low yields. The current efficiencies for hydrocarbon production increased rapidly with rising temperature in the range of 0 to 40°C.[42]

The strong influence of the electrode potential was shown in the presence of tetraalkyl ammonium salts. On glassy carbon electrodes, with 0.1 M aqueous tetramethyl ammonium chloride (pH 8 to 10) as electrolyte, at an electrode potential of -0.9 V vs. SCE, the reduction product of CO_2 was oxalate. At a potential of -1.7 V, the product was glyoxalate. No formate was produced.[43] On graphite and mercury electrodes, with aqueous solutions containing tetramethyl ammonium chloride (0.1 M) at pH 9, CO_2 was reduced in two steps, yielding oxalic acid at -0.9 V, and glyoxylic acid at -1.8 V (vs. SCE). These products were clearly identified by HPLC.[44]

At polycrystalline TiO_2 and TiO_2–Ru electrodes in aqueous 0.5 M KCl, cyclic voltammetry measurements suggested the existence of a strong interaction between the dissolved CO_2 and the hydrated surface of these semiconductor electrodes.[45] Even stronger irreversible adsorption of CO_2 from aqueous solutions or in the presence of water vapor was observed with platinized TiO_2. In this case the formation of two distinct products of CO_2 reduction were indicated by the cyclic voltammetry curves.[46] The effect of the pH of the medium on the shape of the voltammetry curves was the subject of some controversy.[47,48]

An interesting electrode material may be titanium diboride, TiB_2, which is a ceramic polycrystalline material prepared from the powder by a hot press method. It has metallic character. Experiments using it as a cathode for the reduction of carbon dioxide were performed in aqueous media containing 1 M $NaClO_4$. Cyclic voltammetry measurements showed a cathodic reduction peak, the current intensity of which was proportional to the potential sweep rate. This indicated that the electrode reaction proceeds with surface-adsorbed species. The stable reaction products were CO and possibly formic acid.[49,50]

Reduction to Hydrocarbons

The earlier work on the electrochemical reduction of carbon dioxide on metal electrodes searched mainly for the production of formic acid.

More recent work showed that the nature of the electrode metal for carbon dioxide reduction in aqueous electrolytes strongly affected the product composition. Electrode metals could be classified into four main groups: on electrodes of Cd, Hg, In, Sn and Pb, the reduction selectively produced formic acid; on Au, Ag, and Zn, the selectivity was toward carbon monoxide formation; on Cu electrodes, hydrocarbons (mainly methane and ethene), aldehydes, and alcohols were produced; on the other hand, Al, Ga, Pt, Fe, Ni, and Ti have little activity for CO_2 reduction[51-59] (see also Section D). In a more extensive test, of 32 metal electrodes, preferential HCOOH production occurred on the heavy metals in the periodic table groups IIB, IIIB, and IV: Cd, In, Sn, Pb, Tl, Hg, Zn, and Pd. Carbon monoxide was preferentially produced on Ti, Ni, Ag, and Au electrodes. Cu was the only metal that favored the efficient production of hydrocarbons.[60]

Carbon dioxide reduction in CO_2-saturated aqueous $KHCO_3$ (0.05 M) on various metal electrodes was measured at close to 0°C and compared with room temperature electrolysis. In most cases the current efficiencies of product formation were dramatically enhanced at low temperatures, probably because of higher solubility of CO_2, and perhaps also because of the longer lifetime of the adsorbed reaction intermediates. For Ni electrodes, at a potential of -2.2 V vs. SCE, the current efficiencies at room temperature and at 2°C were for CH_4 0.1 and 0.7%, for CO 0.6 and 21%, for C_2H_4 0.01 and 0.07%, for C_2H_6 0.02 and 0.18%, and for HCOOH 0.1 and 13.7%, respectively.[61]

On high-purity copper electrodes in aqueous media, the major products were methane, ethene, ethanol, n-propanol, carbon monoxide, formic acid, and hydrogen. The yield of methane was highest at 0°C. The sum of the Faradaic yields of the carbon dioxide reduction products exceeded 90%. Methanol, formaldehyde, and ethane were not formed. High current yields were achieved only with large overpotentials. The onset potential for methane and ethylene production was -1.5 V vs. SCE. The primary reduction intermediate was proposed to be CO. Carbon monoxide, while weakly adsorbed on the Cu electrode, interfered with cathodic hydrogen production. The adsorbed CO was thus reduced to hydrocarbons and alcohols, with the yields rising with more negative potentials.[12,51-55,62,63] Electrochemical reduction of carbon dioxide to methane was also obtained on Cu-coated glassy carbon and platinum electrodes. In 1 M $KHCO_3$, with a Cu/glassy carbon electrode, the current efficiency for methane formation was 60% at 0°C.[64] The selectivity in the products distribution for CO_2 reduction on Cu electrodes was found to be changed drastically even with small variations in the electrode potentials. In 0.1 M $KHCO_3$ at 298 K, maximal Faradaic efficiencies were 32% at -1.40 V (vs. Ag/AgCl/saturated KCl) for $HCOO^-$, 33% at -1.52 V for CO, 41% at -1.58 V for C_2H_4, and 39% at -1.70 V for CH_4, respectively. For the aldehydes and alcohols, the maximum values were 14% for C_2H_5OH at -1.64

V and at -1.58 V each for CH_3CHO 2%, n-C_3H_7OH 4.5%, and C_2H_5CHO 5%.[58] On Fe electrodes, which were effective for the reduction of CO to hydrocarbons, CO_2 was *not* reduced.[65]

The surface treatment of the Cu electrodes strongly affected the selectivity of the products formed by CO_2 reduction. Rough-surface area electrodes favored high Faradaic yields of CO and $HCOO^-$, while smooth-surface electrodes favored CH_4 and C_2H_4 production.[66] Rough-surface Cu electrodes were more active than smooth ones both in respect of current efficiency and of catalytic electrode life. The electrode deactivation was proposed to be due to adsorbed organic intermediates.[67] The deposition of cadmium on copper electrodes also modified the product selectivity. With increased Cd deposition, the hydrocarbon formation decreased, and the production of CO and $HCOO^-$ increased.[68]

On the above very high-purity Cu electrodes, excellent Faradaic yields of methane were obtained, but the current densities were rather low, at most 10 mA cm^{-2}.[51,52] Increased current densities were achieved using less pure copper (nominally 99.9% pure, with various heavy metal impurities) as the cathode, in 0.5 M $KHCO_3$ as the electrolyte, at 0°C, with a Nafion membrane separating the catholyte from the anolyte. Under continuous CO_2 bubbling through a glass frit, at constant current electrolysis, at a current density of 38 mA cm^{-2}, and with the Cu electrode at a potential of -2.29 V vs. SCE, the Faradaic yield for methane production was 33%.[69]

In order to check if formaldehyde, formic acid, or acetaldehyde may possibly be intermediates during the electrochemical reduction of CO_2 to CH_4 and higher hydrocarbons at Cu electrodes, the electrolysis of these compounds was performed on Cu in both alkaline and acidic aqueous media. The results showed that HCOOH, HCHO, and CH_3CHO were indeed reduced to methane, and may thus be intermediates in the CO_2 reduction. On the other hand, methanol was not reduced to methane under the same conditions, and thus probably is not an intermediate.

On nickel electrodes, the reduction of aqueous CO_2 yielded mainly H_2, with very low yields of CH_4, C_2H_4, C_2H_6, HCOOH, and CO. The difference in the behavior of copper and nickel electrodes was correlated with the very strong adsorption of the intermediate CO on nickel, while CO is only moderately adsorbed on copper. A tentative reaction mechanism on Cu electrodes is the intermediate formation of adsorbed CO as a precursor for the production of hydrocarbons and alcohols. This hypothesis is supported by the similarity in the electroreduction products of CO to that of CO_2 on Cu electrodes. The intermediate CO molecule, which is weakly adsorbed on the copper electrode, is further reduced by hydrogen produced to hydrocarbons and alcohols. On the other hand, on Ni electrodes, CO is very strongly adsorbed, interfering with hydrogen evolution.[62,72]

The dependence of the product distribution on the various metal electrodes was further explained by the different adsorption energies of the CO_2 reduction intermediates.[73,74] These adsorption energies were estimated from the d-band energies of the metals. In group VIII metals, such as Ni, the Fermi levels are situated in the partially filled d-bands. These provide strong covalent σ-bonding to adsorbed CO_2^- anion radicals, which have an unpaired electron in a localized sp^2-hybrid orbital. The strong adsorption energy is thus provided by the σ-bond energy. Therefore, reduction on such metals produces mainly hydrogen. In the group IB metals Cu, Ag, and Au, the Fermi level is situated in the sp-band. The d-band in these metals is slightly below the Fermi level, e.g., 1.5 eV for Cu, and is completely filled. Upon raising the negative potential on the electrode to that of the Fermi level, an electron will be excited from the d-band to the Fermi level, which thus will obtain an unpaired electron, which can form strong covalent σ-bonding with an adsorbed CO_2^- anion radical. The adsorption energy in this case will be the σ-bond energy less the excitation energy. Therefore, with the group IB metals, the adsorption energy is much less than with the group VIII metals. Reduction intermediates, such as CO_2^- will be further reduced to adsorbed CO, CH_2, and HCO radicals, which will readily desorb to free CO, alcohols, and hydrocarbons. In the case of Ag and Au electrodes, the adsorption energy for the intermediate adsorbed CO is very weak and CO is released as the main reduction product. With Cu electrodes, the adsorption energy for the intermediate CO is slightly larger, and thus the lifetime of the adsorbed CO is longer, and it will undergo further reduction to hydrogenated intermediates, which will eventually be released to free hydrocarbons and alcohols. With the group IIB, IIIB, and IVB metals, such as Hg, In, Sn, and Pb, the d-band is deep below the Fermi level, and electronic excitation from the d-band to the Fermi level is impossible. Intermediate CO_2^- radicals can adsorb only extremely weakly on these metals. The weakly adsorbed CO_2^- radicals undergo further reduction by successive H^+ and electron captures, producing HCOOH as the main product.[74]

In the electrochemical reduction of carbon dioxide on Cu electrodes, the selectivity for methane formation was enhanced by the presence of small amounts of methanol. Thus, in an electrolyte of 0.1 M $KHCO_3$, in the absence and presence of 1 mM methanol at an electrode potential of -2.2 V (vs. Hg/Hg_2SO_4), the current efficiencies for CH_4 production were 24 and 47%, respectively. This enhancement of methane formation was at the expense of decreased hydrogen production. The reason for this effect of methanol is not clear.[75]

High current densities, up to 25 mA cm^{-2}, were achieved on copper deposited *in situ* on glassy carbon electrodes, in aqueous 0.5 M $KHCO_3$ saturated with CO_2, operated at -2.0 V vs. SCE. Onset potentials were -1.6 V vs. SCE. Faradaic yields of methane + ethene reached 79%. The mechanism proposed involved adjacent sites of hydrogen atoms and CO_2 adsorbed on copper,[76]

$$H_2O + Cu + e^- \leftrightarrow Cu{-}H_{ads} + OH^- \tag{15}$$

$$Cu + CO_2 \leftrightarrow Cu(CO_2)_{weakly\ ads} \tag{16}$$

followed by the rate-determining step,

$$Cu{-}H_{ads} + Cu(CO_2)_{weakly\ ads} \rightarrow Cu{-}OCHO \tag{17}$$

Also, with Cu deposited on membranes of the solid polymer electrolyte Nafion, carbon dioxide reduction to ethene and ethane (but not to methane) was obtained from gas-phase carbon dioxide even at ambient temperatures. If in addition platinum was deposited on the anodic side of the Nafion membrane, methane was also produced.[77] This gas-phase heterogeneous electrochemical reduction of CO_2 was further tested for a variety of other metals, using the configuration CO_2, M/ Nafion 417/Pt, N_2 + H_2 (90:10), where M was a metal electrocatalyst deposited onto the Nafion polymer electrode. Carbon dioxide flowed through the cathodic compartment, while the nitrogen-hydrogen mixture flowed through the anodic compartment. With M = Ni, Ru, Pb, Pd, Ag, Re, Os, Ir, Pt, and Au, the Faradaic efficiencies for the formation of methane and of higher hydrocarbons were very low.[78]

The mechanism of this CO_2 reduction on copper electrodes was proposed to involve electrochemical splitting of adsorbed CO as the rate-determining step, followed by hydrogenation of surface carbon atoms — in an electrochemical analogue of the Fischer-Tropsch reaction. Methane was produced from carbon dioxide reduction at a 50 times higher rate than from carbon monoxide, which was explained by the approximately 40 times higher solubility of CO_2 than CO in the electrolyte. Since the onset potentials for CH_4 and C_2H_4 formation were the same, -1.5 V vs. SCE, both reactions were probably limited by the same step of CO dissociation,[79]

$$CO_{ads} + H_2O + e^- \leftrightarrow C_{ads} + OH_{ads} + OH^- \tag{18}$$

The black deposits which were always formed on the copper electrodes were shown by XPS and AES (X-ray photon and Auger-electron spectroscopy) to be graphitic carbon, suggesting the following sequence of reactions,[80]

$$CO_2 \rightarrow CO \rightarrow \text{surface bound formyl } \{Cu-HCO\} \rightarrow \tag{19}$$
$$\text{surface carbene } \{Cu=CH_2\} \rightarrow \text{hydrocarbons}$$

Surface carbenes (methylenes) had previously been proposed as the intermediates in the Fischer-Tropsch synthesis of hydrocarbons.[80] The similarity of CO_2 reduction mechanisms on metal electrodes to the Fischer-Tropsch hydrogenation of CO_2 on these metals seems indicated by the similarity in the distribution of the reduction products, which include CO, alkanes, alkenes, aldehydes, and alcohols. This product distribution may be represented by the relationship,

$$\ln(m_p/P) = \ln(\ln^2\alpha) + P(\ln\alpha) \tag{20}$$

where P = the number of carbon atoms in a given hydrocarbon, m_p = the weight fraction of the oligomer having P carbon atoms, and α = the probability of chain growth. The linearity of plots of $\ln(m_p/P)$ vs. P for the electroreduction of CO_2

on various metals was proposed as a criterion for a Fischer-Tropsch mechanism, presumably involving intermediate adsorbed CH_x species.[81,82]

On electroplated ruthenium cathodes in aqueous Na_2SO_4 saturated with carbon dioxide, CO_2 reduction resulted in formation of both methane and methanol, with current yields of 30 and 35%. However, the current densities were less than 1 mA cm^{-2}. The proposed mechanism included hydrogenation of a surface carbon species as the rate-determining step.[83] Ruthenium is known as an active catalyst for CO_2 methanation and Fischer-Tropsch-type gas-solid reactions. The suggested mechanism is hydrogenation of a surface carbon species as the rate-limiting step,[84]

$$H_{ads} + CH_\chi \rightarrow CH_{\chi+1} \text{ where } \chi = 0 \text{ to } 3 \tag{21}$$

An interesting approach to energy storage has been that of storing hydrogen in copper-containing palladium metal electrodes. On such Cu-Pd/H_2 cathodes, in carbon dioxide-saturated 0.1 M $KHCO_3$, at a potential of -1.7 V (vs. SCE), the Faradaic efficiencies for production of CO, CH_4, HCOOH, and H_2 were 0.76, 24.0, 54.0, and 29.6%, respectively.[85]

The mechanism of reduction of carbon dioxide on copper electrodes to methane and ethene was also studied by on-line electrochemical mass spectrometry with rotating electrodes. Electrodeposited copper on glassy carbon was found to be a better catalyst for methane formation than bulk copper. The onset potential for hydrocarbon formation was -1.7 V vs. SCE, detected both by voltammetry and mass spectrometry.[86]

Reduction to Methanol and Ethanol

There has been considerable effort to produce methanol by direct electrochemical reduction of carbon dioxide. This has been successful with a number of electrode systems, sometimes with high Faradaic efficiencies, but the current densities have often been disappointingly low, usually less than 1 mA cm^{-2}.

Using n-GaAs single-crystal electrodes (111 As face), in CO_2-saturated Na_2SO_4 solution, methanol was produced with 100% Faradaic yield, at electrode potentials of -1.2 to -1.4 V (SCE), but at current densities of only 0.2 mA cm^{-2}. Much lower yields were observed at (110) and (100) faces, and using p-GaAs and p-InP as photocathodes.[87,88]

However, on repeating the reduction of CO_2 on heavily doped p-GaAs, using the (111 As) face, the conversion to methanol occurred even at open circuit in the dark, at rates reaching 0.1 μmol h^{-1} cm^{-2}. An even larger rate was found with a moderately doped n-GaAs crystal. On the (Ga) and (100) faces, methanol production was much lower. Closed-circuit electrolysis resulted mainly in formic acid production. The explanation given for the open circuit effect was the corrosion (dissolution) of the GaAs in carbonic acid.[89]

On molybdenum electrodes in CO_2-saturated acidic aqueous solutions (0.2 M Na_2SO_4, pH 4.2 or 0.05 M H_2SO_4) at -0.7 to -0.8 V vs. SCE, methanol was formed as the major carbon dioxide reduction product, with Faradaic efficiencies of more than 50%, and current densities of up to 0.6 mA cm^{-2}. Minor products included CO and CH$_4$.[90]

On electrodes of ruthenium fixed on polyhydroquinone/benzoquinone (prepared by electropolymerization of mercaptohydroquinone), supported on a glassy carbon electrode, the onset potential for CO$_2$ reduction was as low as -0.5 V (vs. SCE). At -0.7 V (vs. SCE), in 0.2 M Na$_2$SO$_4$, the current efficiency for methanol production attained almost 100%, but at a current density of only 0.2 mA cm^{-2}. Under illumination with visible light, the efficiency for methanol production even exceeded 100%,[91] an effect which may possibly be understood by the corrosion of the electrode.

On Sr-doped TiO$_2$ electrodes, at -1.7 V (vs. SCE), the Faradaic yield for methanol production reached 3.1%, while the yield on the undoped electrodes was only 1.3%.[92]

On RuO$_2$ + TiO$_2$ (35:65) coated on titanium foil, at -0.1 V (vs. Hg$_2$SO$_4$) in 0.05 M H$_2$SO$_4$ (pH 1.2), the reduction of CO$_2$ to methanol occurred with 24% current efficiency, and at a current density of 0.52 mA cm^{-2}. The other product was formic acid (2%). Tafel plots of log(current density) vs. electrode potential gave Tafel slopes of 180 to 240 mV.[93] These were similar to those previously observed at the higher current densities for CO$_2$ reduction on Hg electrodes, which had been assigned to the first electron transfer reaction as the rate-determining step.[20] The mechanism proposed involved as a primary fast step the partial reduction of the metal oxide surface,[93]

$$MeO_2 + H^+ + e^- \rightarrow MeOOH \tag{22}$$

followed by a rate-determining step of chemisorption of the CO$_2$ molecule, with formation of a carbonate intermediate.[93] With mixed TiO$_2$ + RuO$_2$ electrodes (ratio 1:3, corresponding to an oxide composition of Ru$_3$TiO$_8$, deposited by decomposition of the mixed halides in air at 450°C on Ti plates), the electrolysis of aqueous 0.5 M KHCO$_3$ was performed at a potential of -0.90 to -0.95 vs. Ag/AgCl and a current density of 5 mA cm^{-2}. In the presence of CO$_2$, the current efficiency for CO$_2$ reduction reached a maximal value at potentials just before the steep onset of H$_2$ evolution. The rate-limiting step was proposed to be a surface recombination of adsorbed hydrogen with CO$_2$,[94]

$$CO_2 + H_{ad} + e^- = COOH_{ad}^- \tag{23}$$

Electrodeposition of small amounts of copper on these TiO$_2$ + RuO$_2$ electrodes caused very much enhanced yields of methanol, ethanol, and formic acid. Copper deposition shifted the current-potential curves for H$_2$ evolution to more negative potentials, while the CO$_2$ reduction efficiency increased.[94]

Since the adsorption of carbon dioxide on the electrode surface is an essential prerequisite for the electron transfer which is the primary reaction step, some effort has been made to correlate the electrochemical activity of different electrode materials with their bond properties. The adsorption properties of several metals and metal oxides for carbon dioxide were determined by pseudocapacity measurements. The surfaces of electrodes with mixed oxides, such as RuO_2-TiO_2 and $RuO_2-Co_3O_4-SnO_2-TiO_2$ which showed good electrocatalytic activity for CO_2 reduction to methanol, also showed increased pseudocapacity under CO_2 saturation (compared with that under N_2). On the other hand, on electrodes with mixed oxides such as IrO_2-TiO_2 or pure IrO_2, which were inactive for CO_2 reduction, there was no pseudocapacity in CO_2-purged solutions, presumably because the surfaces were totally obstructed by the strong absorption of CO_2. The electrochemical behavior was correlated with the d-bond character of the different metals. High pseudocapacity for electrodes in CO_2-purged solutions occurred with the metals Cu, Ni, and Au.[95]

On Cu electrodes, the product distribution depended on the electrode potential. In 0.1 M $KHCO_3$ at 298 K, ethanol was produced at maximal Faradaic efficiency of about 14% at a potential of -1.65 V (vs. Ag/AgCl saturated KCl). Other products in smaller yields were CH_3CHO, C_2H_5CHO, and $n-C_3H_7OH$.[96]

At electrodes of oxidized copper, prepared by anodizing or thermally air oxidizing copper foil, carbon dioxide reduction to methanol was observed with onset potentials as low as -0.4 V (SCE). Surprisingly, the Faradaic yields for methanol production considerably exceeded 100%, indicating that the reaction is not a simple six-electron reduction. The highest rate of methanol production, 1×10^{-4} mol cm^{-2} h^{-1}, was obtained with anodized copper in 0.5 M $KHCO_3$ at pH = 7.6 and -1.9 V (SCE). The mechanism proposed to account for the very high Faradaic yields (or "open-circuit" CO_2 reduction) involves chemical reduction steps up to HCO_{ads}, followed by three hydrogenation steps to methanol.[97]

Carbon dioxide was reduced to methanol at a Prussian blue, $[KFe(III)Fe(II)(CN)_6]$, coated platinum cathode, with an illuminated n-CdS photoanode as energy source, and pentacyanoferrate as mediator, in a medium containing 0.02 M methanol. The current efficiency for methanol production was reported to be 87%. Other metal complexes were also very effective as mediators, such as 1-nitroso-2-naphthol 3,5-disulfonic acid. The mechanism proposed involves a labile ligand in the metal complex, which is the active site in the catalytic process. A coordination bond is formed between the primary alcohol and the neutral metal atom. Carbon dioxide inserts into this bond, producing a formate-like intermediate, which is reduced by the neighboring surface-bound Fe(II) complex, forming methanol. In the absence of a primary alcohol in the medium, the CO_2 reduction product was HCOOH.[98-100] Instead of a photoelectrode or of an external voltage source, a hydrogen fuel cell of Pt-gauze could be used as the anode, at which H_2 was oxidized. CO_2 was reduced by the above metal complex at a Prussian-blue-coated Pt cathode. At short circuit, methanol formation was maximal, reaching 93% current efficiency, but the current density was only about 1 μA cm^{-2}.[101]

Homogeneous catalysis of carbon dioxide electroreduction was achieved using a system consisting of cobalt(II)-2-nitroso-1-naphthol 4-sulfuric acid complex with methanol. The electrode was modified by a surface layer of a water-insoluble complex, Everitt's salt, which is the reduced form of Prussian Blue, $K_3Fe(III)[Fe(II)(CN)_6]$. With this mediated electrochemical system, CO_2 was reduced to methanol in the potential range -0.15 to -0.35 V (vs. SCE), according to the reaction,[102]

$$CO_2 + 6H^+ + 6e^- \rightarrow CH_3OH + H_2O \qquad (24)$$

Alloy Electrodes

Alloy electrodes were developed in order to improve the selectivity and efficiency of alcohol production. On metals with high overpotential for hydrogen production, such as cadmium, indium, lead, mercury, silver, and copper, the current efficiency and product selectivity for carbon dioxide reduction to carbon monoxide, formic acid or methane, and ethane is good, but the overpotentials are high and the energy efficiency is low. On the other hand, on metals with low overpotentials for hydrogen generation, such as from group VIII of the periodic table, both the selectivity and the energy efficiency for carbon dioxide reduction are poor. Improvements in the selectivity and energy efficiency of CO_2 reduction was achieved by using alloy electrodes. These depend on the introduction of low hydrogen-overpotential metals such as group VIII metals on the surface of high hydrogen-overpotential metals (such as the group IB or IIB metals). These alloy electrodes enabled catalytic reduction of surface-adsorbed carbon dioxide or reduced carbon intermediates by surface-adsorbed hydrogen atoms. On Cu-Ni alloys (atomic ratio 90.5/9.5), the formation of methanol occurred at almost the reversible potential (onset potential was -0.38 V vs. SHE), with a Faradaic efficiency of about 10% at a potential of -0.65 V (vs. SHE). Alloy electrodes were prepared by electro-plating the appropriate salt mixtures onto gold flag electrodes, and were tested in CO_2-saturated aqueous 0.05 M $KHCO_3$.[103] For production of CO and formic acid, the alloy electrodes Cu–Sn, Cu–Pb, and Cu–Zn provided low overpotentials and hence high energy efficiency. However, the current densities on these electrodes were low. For methanol and formic acid production, the partial current densities on the Cu–Ni alloy were only 0.008 and 0.28 mA cm^{-2}, respectively.[103]

With Cu–Ag (2:3 atomic ratio) alloy electrodes, prepared by electron-beam evaporation of a silver layer (about 20 to 30 nm thick) on a clean copper plate (99.999% pure), and annealing under vacuum at 300 to 400°C for about 20 min, much-enhanced production of C_2H_4 was obtained, while the production of CH_4 was slightly lowered. The explanation given for the improved current efficiency for ethylene on the alloy electrodes is that, on the Ag grains on the surface of the alloy, CO molecules will be produced (as on pure silver electrodes). In their vicinity, on the Cu grains, chemisorbed H, CO_2^-, CO, HCO, and CH_2 radicals will be formed (as on pure Cu electrodes). Since these grains are presumably randomly distributed, the probability of interaction between the CO molecules on the Ag grains with the H atoms on the Cu grains becomes enhanced, leading to higher production of CH_2 radicals, which dimerize to ethylene.[73]

Mediation by Macrocycles and Metal Complexes

Direct electrochemical reduction of carbon dioxide requires on most electrode materials a considerable overvoltage, thus decreasing the energy conversion efficiency. Considerable decreases in overvoltage can be attained by using transition metal complexes as electron-transfer mediators.[107]

Metalloporphyrins in aqueous solutions were found to catalyze the electroreduction of carbon dioxide to formic acid. The complexes investigated included tetraphenyl-porphine sulfonates and *meso*- tetracarboxyphenyl-porphyrins. Cobalt, copper, and iron porphyrins were tested in alkaline media (pH 8 to 10), with mercury pool cathodes. Only the cobalt porphyrins catalyzed the reduction of carbon dioxide.[108]

Carbon dioxide at pressures of 4 to 22 atm was electroreduced in aqueous phosphate buffer solutions ($0.5\ M\ NaH_2PO_4$ + NaOH) containing cobalt tetrakis (4-trimethylammonio-phenyl)-porphyrin as mediator, using In, Sn, Pb, and Pb–Hg (lead amalgam) electrodes. Maximal current efficiency for CO production was in the order Pb–Hg > Pb > In > Sn, while, for HCOOH production, the order of electrode activity was Pb > Pb–Hg > Sn > In. The optimal potential for CO_2 reduction to both CO and HCOOH was -1.2 V vs. SCE. Current efficiencies reached >90% with Pb and Pb-Hg electrodes.[109]

Optically transparent thin-layer electrodes (OTTLE) were used to study the electrocatalytic reduction of CO_2, using tetrakis (4-trimethylanilino) porphinato cobalt^{2+} tetraiodide (CoTMAPI) as catalyst. Applying a spectroelectrochemical technique, the electrode potentials were kept at $+0.6$, -0.3, and -1.0 V. Under these conditions, the UV absorption spectra of the mono-, di-, and trivalent ions were obtained. Controlled potential electrolysis carried out at an electrode potential of -1.0 V in a CO_2-saturated solution of CoTMAPI in the presence of imidazole indicated a catalytic cycle for CO_2 electroreduction.[110]

Various other metallo macrocycles in aqueous or organic solvents were found to catalyze the reduction of carbon dioxide. Nickel-cyclam dichloride (cyclam = 1,4,8,11-tetraazatetradecane),

mediated the selective reduction of CO_2 to CO, according to the overall reaction,

$$CO_2 + 2e^- + 2H^+ \rightarrow CO + H_2O \tag{25}$$

At a potential of -1.0 V (vs. NHE), in 0.1 M aqueous KNO_3, the current efficiency for CO production reached 96%. Also, in dimethylformaldehyde as solvent, formic acid was produced in 75% current efficiency. The reduction of CO_2 was proposed to involve a Ni(I)carbonyl complex as intermediate.[111-113] The mechanistic steps for Ni-cyclam-mediated CO_2 reduction on Hg electrodes were proposed to include (1) adsorption of the reduced form of the mediator on the electrode surface, (2) coordination of CO_2 on the electrode surface, and (3) further electron transfer, from the electrode surface through the catalyst to the CO_2 group.[114]

In the electroreduction of CO_2 to CO on Hg electrodes mediated by Ni(cyclam)$^{2+}$, the catalytically active species was suggested to be Ni(cyclam)$^+$ adsorbed on the mercury. A detailed cyclic voltammetry study indicated that Ni(cyclam)$^{2+}$ was only weakly adsorbed at mercury electrodes, and only over a limited potential range, while Ni(cyclam)$^+$ was strongly adsorbed, and over a wide potential range. In its activity as electrocatalyst for CO_2 reduction, an altered configuration of the cyclam ligand may possibly play a role.[115]

Monolayers of two amphiphilic nickel(II) complexes with long-alkyl-substituted cyclam derivatives were deposited onto glassy carbon disc electrodes, using the Langmuir-Blodgett technique. The complexes were nickel tetrakis (N-hexadecyl) cyclam and nickel N-hexadecyl cyclam. The electrocatalytic activities of these electrodes for the reduction of carbon dioxide were examined by cyclic voltammetry. In aqueous solutions at pH 4.5, the cyclic voltammogram with the monolayer of Ni tetrakis (N-hexadecyl) cyclam or Ni-N-hexadecyl cyclam-coated electrodes indicated strong rising catalytic currents for the reduction of carbon dioxide at -1.30 V (vs. SCE). This potential was only slightly different from that observed for nickel cyclam in solution, measured on a hanging drop mercury electrode.[116,117] Since the Ni-N-hexadecylcyclam monolayer was found to be rather unstable, possibly because of appreciable solubility, the Ni(II) complex of a cyclam with a much longer alkyl chain was tested, Ni(II)-N-docosyl-cyclam [docosyl $= CH_3(CH_2)_{20}CH_2-$]. This cyclam, when deposited on a glassy carbon electrode by the Langmuir-Blodgett technique, provided the electrocatalytic reduction of CO_2 already at an onset potential of only -1.25 V vs. SCE. The first catalyst layer on the electrode was more effective for CO_2 reduction than the subsequent layers. In this first monolayer, the cyclam head group was oriented toward the hydrophilic electrode surface.[118]

Using a nonwetting porous Hg-Au (amalgamated gold-mesh) electrode at the polytetrafluoroethylene membrane inlet system of a differential electrochemical mass spectrometer, the electroreduction of CO_2 in an aqueous solution of Ni(II)-cyclam was studied during cyclic voltammetry. The appearance of the cathodic peak at -1.15 V (SCE) coincided with the mass peaks of CO and H_2.[119-121]

Carbon electrodes (either pyrolytic graphite or carbon cloth) coated with cobalt phthalocyanine were used in CO_2-saturated aqueous citrate buffer solutions (pH

5), producing CO and H_2 with about 60 and 30% current efficiency, at -1.0 V (vs. SSCE), with a current density of about 1 mA cm^{-2}. The turnover number (moles of CO produced per mole of catalyst) exceeded 10^5.[122] However, in another study, with CO_2-saturated acid solutions (pH 3 to 7), on glassy carbon electrodes coated with cobalt or nickel phthalocyanines, the main product was formic acid, obtained with an overpotential which was smaller by 200 mV than with bare metal electrodes. Current densities reached up to 10 mA cm^{-2}. At lower pH values, methanol was also produced, with a current efficiency of up to 5%.[123] On graphite electrodes impregnated with Co or Ni phthalocyanines, with aqueous tetraalkyl ammonium salts as electrolytes, the products identified were oxalic and glycolic acids. Formic acid was not produced. The current-potential curve showed a cathodic peak for CO_2 reduction, the height of which was proportional to the square root of the sweep rate. This indicated that the diffusion of carbon dioxide was the rate-limiting step.[124]

The electrocatalytic activity of the metal phthalocyanines has been explained by the observation that these complexes are reduced to their dinegative states at the potentials required for carbon dioxide reduction.[125] It was suggested that the Co and Ni phthalocyanines in their dinegative states have an excess of ligand π-electrons.[126] This may play a role in their electrocatalysis of CO_2 reduction. An interpretation of the differences in CO_2 reduction products obtained with the different metal phthalocyanines was proposed in terms of the theory of Taube, based on the LCAO-MO Hückel method.[127,128]

A promising method for the electrocatalysis of CO_2 reduction is the plasma-assisted deposition of metal phthalocyanine thin films on glassy carbon electrodes. The technique involved vacuum deposition using an intermittent plasma in a gas mixture of argon and hydrogen. The films were uniform and about 400 Å thick. The electrocatalytic activity of the electrodes was tested in aqueous 0.5 M Na_2SO_4 + $NaHCO_3$ (pH 6.65) by measuring current potential curves with rotating discs of these metal-phthalocyanine-coated electrodes. The relative activity depended on the metal of the complexes, in the order Co $\gg$ Ni $>$ Fe $\sim$ Mg $\sim$ Mn $\sim$ Zn phthalocyanines.[129]

An improved electroreduction of CO_2 to CO with metalloporphyrins as catalysts was achieved by immobilizing water-insoluble cobalt(II)-tetraphenyl-porphyrin (Co^{II}tpp) on glassy carbon electrodes, which had been treated with 4-aminopyridine. On these electrodes, at -1.2 V (vs. SCE), in CO_2-saturated aqueous solutions (phosphate buffer, pH 6.86), carbon dioxide reduction yielded as only products CO and H_2, with current efficiencies of about 50% each. This electroreduction of CO_2 to CO occurred at potentials which were 100 mV more positive than those observed with the water-soluble Co^{II} porphyrins. The turnover number of the immobilized catalyst for CO formation exceeded 10^7.[130,131]

Since bipyridine halotricarbonyl complexes of rhenium(I), [*fac*-Re(I)(bpy)(CO)$_3$Cl] (where bpy = 2.2-bipyridine), had been shown to be excellent catalysts for the electroreduction of CO_2 in nonaqueous solvents (see Section II.C), the incorporation of such complexes into water-insoluble polymer coatings on metal electrodes enabled their use in aqueous media. Such coatings with poly(pyrrole) films containing the Re(bpy)(CO)$_3$Cl system on platinum electrodes were found to be efficient electrocatalysts for the reduction of CO$_2$ to CO and CO$_3^{2-}$ as the only products in high current yields.[132,133] Incorporation of electron-withdrawing carboxy ester groups in the bpy ligand of these Re(bpy)(CO)$_3$Cl complexes, both in solution and in polymeric form, strongly stabilized the initial oxidation state Re(I) of the metal center, and decreased the electrocatalytic activity toward CO$_2$ reduction.[134]

Metal complexes incorporated into hydrophobic polymer films coated on electrodes were used to achieve electrocatalytic reduction of carbon dioxide. With glassy carbon electrodes coated with a Nafion membrane, and on this complexes of Ru, Co, or Re, such as Co(terpy)$_2^{2+}$, at -1.55 V (SCE), in aqueous phosphate buffer or KHCO$_3$, formic acid was produced in about 10% Faradaic efficiency.[135]

The electroreduction of CO$_2$ to C$_1$ to C$_3$ hydrocarbons was observed in the presence of a catalytic system which involved pyrocatechol complexes of Ti(III) and Mo(III). The electrolysis was performed in a two-compartment cell, with a ground-glass joint as separator, with a mercury-pool cathode at -1.55 V vs. SCE, and a catholyte prepared from an aqueous solution of 0.5 M pyrocatechol, 0.05 M TiCl$_3$, and 5 mM Na$_2$MO$_4$. Products observed were methane, ethane, ethylene, and C$_6$ hydrocarbons (in the ratio 4.2:0.45:0.16:0.12), with a total current efficiency of up to 0.2%, as well as copious evolution of hydrogen. The reduction of CO$_2$ was postulated to occur within the coordination sphere of Mo(III), possibly through carbene complexes such as Mo = CH$_2$. The Ti(III)-pyrocatechol complex presumably acted as electron mediator.[136]

Gas Diffusion Electrodes

The challenge of highly efficient electroreduction of carbon dioxide requires the achievement of both high current densities and low overpotentials. The rate of CO$_2$ electroreduction in liquid phase electrolysis is often limited by mass transfer. Gas diffusion electrodes enable a considerable enhancement of mass transfer to the triple gas-electrolyte-solid electrode boundary.[137] Very high current densities of CO$_2$ reduction to formic acid have been obtained during the last few years by using metal-impregnated gas diffusion electrodes. On such electrodes, with lead-impregnated polytetrafluoroethylene (PTFE)-bonded carbon operated at -1.8 V vs. SCE in aqueous solutions (pH 2), HCOOH was produced with nearly 100% current efficiency, at the remarkably high current density of 115 mA/cm^2. The electrocatalytic activity for carbon dioxide reduction to formic acid was in the order Pb > In > Sn.[138]

Since electrodes of pure copper had been shown to favor the production of hydrocarbons in the electroreduction of carbon dioxide in aqueous solutions, an effort was made to optimize the performance of gas-diffusion electrodes loaded with copper. Carbon dioxide reduction on a finely divided copper electrode was achieved, using a proton-conducting solid polymer structure. The Cu/Nafion electrodes were fabricated by an electroless plating method. The gas phase electro-chemical reduction of CO_2 to CH_4 and C_2H_4 was optimal in 1 mM H_2SO_4. At a potential of -2.0 V vs. SCE, the steady-state Faradaic efficiency reached about 20%.[139]

Gas-diffusion electrodes were also fabricated by bonding together a semihy-drophilic reaction layer (containing both hydrophilic and hydrophobic carbon black, PTFE, and Cu powder of 5 N purity) and a hydrophobic gas-supply layer (hydrophobic carbon black, PTFE, and 20-mesh Cu gauze as current collector). With such electrodes in 0.5 M K_2SO_4 supplied with CO_2, the products of reduction were CH_4, C_2H_4, C_2H_5OH, CO, $HCOO^-$, and H_2, formed optimally at a potential of -1.45 V vs. Ag/AgCl, with a total Faradaic efficiency of 70%. The current densities were about 100 times higher than those on copper plate electrodes.[140]

At gas-diffusion electrodes impregnated with cobalt(II) phthalocyanine, high rates of reduction of CO_2 to CO were also demonstrated, reaching a current density of 137 mA cm^{-2} at a potential of -2.2 V (vs. SCE). On the other hand, on electrodes impregnated with manganese, copper, or zinc phthalocyanine, formic acid was produced at low rates, and no carbon monoxide was formed. The unique effectiveness of cobalt(II) phthalocyanine was explained by a mechanism involv-ing as a primary step the electrochemical reduction of cobalt(II) to cobalt(I).[141] The selective reduction of CO_2 to CO was confirmed on carbon gas diffusion electrodes promoted by cobalt phthalocyanine. At a potential of -4.4 V vs. Ag/AgCl, current densities of up to 80 mA cm^{-2} were attained for CO_2 reduction to CO, with Faradaic yields of up to 97%, using cobalt phthalocyanine (9%) sup-ported on a carbon gas-diffusion electrode, which was prepared from carbon black and Teflon (50%) and supported on carbon paper. Co-phthalocyanine was de-posited onto the electrode from its solution in dimethylformamide.[137]

With gas-diffusion electrodes modified by metal phthalocyanines, operated at current densities of up to 100 mA/cm^2, the product distribution depended on the central atom of the phthalocyanine: for Cu, Ga, and Ti, the main product was CH_4, with current yields of up to 40%, while for Co, Ni, Fe and Pd the preferred product was CO, and for Sn, Pb, In and Al the major product was HCOOH. Such gas diffusion electrodes were prepared from carbon black, Teflon, and metal phthalocyanines.[142]

Remarkably high production rates of alcohols were achieved by using gas diffusion electrodes based on Teflon-bonded $La_{0.9}Sr_{0.1}CuO_3$, in 0.5 M KOH, with a separator and Pt counterelectrode. At a current density of 180 mA cm^{-2}, the Faradaic yields of ethanol and n-propanol were 30.7 and 10% after 1.25 h of electrolysis. If the copper component in the above perovskite material was replaced by Ni, Co, Fe, Mn, or Pd, the resulting electrodes were inactive for carbon dioxide reduction to alcohols.[82]

In order to identify the intermediates in the electrochemical reduction of carbon dioxide, the technique of Raman spectro-electrochemistry was applied. *In situ* studies of lead-impregnated PTFE-bonded carbon gas-diffusion electrodes re-vealed that during polarization of the electrode surface weak Raman bands at

2824 and 2915 cm^{-1} appeared, assigned to free HCOO$^-$ and HCOOH perturbed at the electrode surface.[143]

By comparision with other electrode systems, gas-diffusion electrodes seem to offer the most practical approach to the efficient reduction of carbon dioxide, because of the high current densities achieved. An essential further target with the gas-diffusion electrodes will be to lower the overpotential required, and preferably to achieve direct CO_2 reduction to alcohols.

REDUCTION IN ORGANIC SOLVENTS

While electron capture by carbon dioxide is generally considered to be the primary step of electroreduction both in aqueous and organic media, subsequent reactions account for the preferred production of formic acid in water and of oxalic acid and carbon monoxide in aprotic solvents,[143-146]

$$CO_2 + e^- \rightarrow \cdot CO_2^- \ (E^\circ = -2.21 \text{ V vs. SCE}) \tag{26}$$

$$2 \cdot CO_2^- \rightarrow C_2O_4^{2-} \ k = 3.2 \times 10^7 \text{ M}^{-1} \text{ s}^{-1} \tag{27}$$

$$2 \cdot CO_2^- \rightarrow O{=}C{-}O(C{=}O){-}O^- \tag{28}$$

$$O{=}C{-}O(C{=}O){-}O^- \rightarrow CO_3^{2-} + CO \ k = 3.2 \times 10^3 \text{ M}^{-1} \text{ s}^{-1} \tag{29}$$

In dry dimethyl formamide solution, the rate constant of electron transfer to CO_2 was shown to be 6×10^3 cm s^{-1}, the transfer coefficient was 0.4, and the second-order rate constant of CO_2^- deactivation was 3.2×10^7 M^{-1} s^{-1}, as shown in Equation 27 above.[147]

Direct Reduction

A comparison of the solvents water, dimethyl sulfoxide (DMSO), acetonitrile (MeCN), propylene carbonate (PC), and dimethyl formamide (DMF) for the reduction of carbon dioxide on electrodes of glassy carbon, mercury, platinum, gold, and lead was performed by measuring the voltammetric reduction waves, with tetramethyl ammonium bromide as the supporting electrolyte. The peak potentials for CO_2 reduction on Hg electrodes in water, MeCN, DMSO, PC, and DMF were -2.16, -2.76, -2.40, -2.59, and -2.21 V (vs. SCE). From an analysis of the voltammetric waves, the apparent number of electrons involved in the reduction of CO_2 was derived, resulting in $n_{app} = 2$ in aqueous solutions on glassy carbon electrodes, $n_{app} = 1$ on Hg and Au electrodes in DMSO solutions, but $n_{app} = 2$ on Pt electrodes in DMSO. In MeCN solutions, the observed n_{app} values with Hg and Pt electrodes were 1 and 2, respectively. The results indicate different reaction mechanisms on the various metal electrodes.[148]

In aprotic organic solvents such as acetonitrile and dimethylformamide the main carbon dioxide reduction products were usually oxalic acid and carbon monoxide. Metal electrodes which strongly chemisorb carbon dioxide, such as platinum, enhanced the production of CO,

$$2CO_2 + 2e^- \rightarrow CO + CO_3^{2-} \tag{30}$$

The mechanism of the electrochemical reduction of carbon dioxide on lead electrodes in acetonitrile and in propylene carbonate solvents was shown by modulated specular electroreflectance spectroscopy to involve two intermediate radical species. These were detected by their strong absorption peaks at 285 and at 330 nm in propylene carbonate solution. In acetonitrile, these peaks were blue shifted to 270 and 315 nm, respectively. The absorption peaks were assigned to the radical anion CO_2^- and the adduct of this radical anion with a neutral CO_2 molecule, $(CO_2)_2^-$,

$$CO_2^- + CO_2 \rightarrow (CO_2)_2^- \tag{31}$$

As noted above (Section I.A), in aqueous media only the single peak of CO_2^- at 250 nm was observed.[18]

A mechanistic study of the electroreduction of carbon dioxide in aprotic solvents used photoemission measurements and steady-state (dark) polarization curves at Hg, Pb, Sn, In, and Pt electrodes. In acetonitrile solutions, the overpotentials of CO_2 reduction increased in the order In < Sn < Pb < Hg. This considerable sensitivity of the overpotential to the electrode material indicated that the rate-limiting step of CO_2 electroreduction in aprotic solvents involves species adsorbed on the electrode surface. For all these metals, except platinum, plots of the steady-state currents against potential consisted of two Tafel regions. In the first Tafel region, with a slope of 140 to 180 mV, the rate-limiting step was proposed to be the reaction of the CO_2^- radical anion (formed in a prior fast electron-transfer step) with a CO_2 molecule to form the $(CO_2)_2^-$ anion radical,[149,150]

$$CO_2 + e^- \rightarrow CO_2^- \tag{32}$$

$$CO_2^-(ads) + CO_2(ads) \rightarrow (CO_2)_2^- \tag{33}$$

$$(CO_2)_2^-(ads) + e^- \rightarrow C_2O_4^{2-} \tag{34}$$

In the second Tafel region, with a slope of 600 to 700 mV, the initial electron capture by CO_2 to form the CO_2^- radical anion became the rate-determining step, just as in aqueous solutions. On platinum electrodes, the polarization curve was flat — there appeared only a single Tafel slope, of 600 to 700 mV, corresponding to the second Tafel region of the other metals.[149,150]

Using Hg or Cr–Ni–Mo steel (18:10:2%) cathodes, on which CO_2 and its reduction products are only weakly absorbed, the Faradaic yield of oxalic acid was 61%, at a current density of 6 mA cm^{-2} The reaction was explained by the high concentration near the cathode of CO_2^- radical anions, the dimerization of which resulted in oxalate production,[26]

$$2CO_2 + 2e^- \rightarrow (COO)_2^{2-} \qquad (35)$$

The mechanism of electroreduction of carbon dioxide to oxalate on platinum electrodes in acetonitrile solution was studied using infrared spectroelectrochemistry. *In situ* FTIR (Fourier transform infrared) reflectance spectra of the thin electrolyte layer between the cell window and the electrode were measured. Oxalate was formed in at least two different solvation states.[151]

On comparing various metals as electrodes for carbon dioxide reduction in nonaqueous solutions, the product composition was found to depend on the nature of the metal. Oxalic acid was predominantly produced on Hg, Tl, and Pb, while carbon monoxide was the preferred product on Cu, Zn, In, Sn, and Au.[56,57,96,152,153] With a high-pressure CO_2 electrolysis cell, using indium, tin, or lead cathodes, with tetraalkyl ammonium salts as electrolytes, high efficiency of formic acid production was attained. This product reached a concentration of up to 1 M. Also, propionic, n-butyric, and oxalic acids were formed in small yields.[152]

A practical process was developed by applying elevated pressures of carbon dioxide. Using stainless steel cathodes, sacrificial aluminum anodes, in a diaphragm-less system containing 0.1 M tetrabutyl ammonium bromide in dimethylformamide as electrolyte, with CO_2 at pressures of 1.4 to 31 atm, aluminum oxalate was produced with current yields of up to 38%, as well as CO and carbonate. The anodic and cathodic reactions were represented by,[154–156]

$$Al \rightarrow Al^{3+} + 3e^- \qquad (36)$$

$$2CO_2 + 2e^- \rightarrow {}^-OOC-COO^- \qquad (37)$$

$$2CO_2 + 2e^- \rightarrow CO + CO_3^{2-} \qquad (38)$$

A semi-pilot plant process for carbon dioxide reduction to oxalic acid was developed, using sacrificial zinc anodes and stainless steel cathodes in a single compartment cell, with acetonitrile solutions of tetrabutyl ammonium perchlorate as the electrolyte. At current densities of about 5 to 10 mA cm^{-2}, Faradaic efficiencies of more than 90% were achieved. The product was insoluble zinc oxalate, which was collected by filtration. After hydrolysis of the zinc oxalate, the zinc could be recovered as metal by electrolysis, to be recycled to the carbon dioxide reduction stage. The whole process was performed without any waste products.[157]

A variety of electrolysis cell types were tested in an effort to increase the current densities and current yields for oxalic acid production in the electrolysis of carbon dioxide. Lead, lead amalgam, stainless steel, cadmium, aluminium, copper, or zinc cathodes were used, in 0.18 M tetraethylammonium bromide in DMF saturated with CO_2, separated by a cationic membrane from graphite anodes in 1 M NaCl, at atmospheric pressure. The highest current yield for oxalic acid production, 51%, was obtained with the Pb-amalgam electrode, at a current density of 10 mA cm^{-2}. Raising the current density to 20 mA cm^{-2} caused a decline in current yield to 29%. Also, at the higher current density, in DMF solution, a black deposit was formed on Pb or Pb-amalgam electrodes. With the other electrode materials, the current yields were much lower. Higher production rates of oxalic acid were achieved using a high-pressure cell. With a lead cathode, at a CO_2 pressure of 690 kN m^{-2} (6.8 atm), the current density reached 50 mA cm^{-2} and the current yield was 44%, but the process was plagued by attack on the lead electrode, as well as by the decomposition of DMF to dimethylamine. Stainless steel electrodes were more stable, and at a current density of 20 mA cm^{-2} gave a current yield of 36%.[158]

Relatively high current densities, of up to 50 mA/cm^2 for the reduction of CO_2 to CO have been achieved by using freshly deposited metal cathodes (best results with Cd, Sn, or Ag) and a sacrificial Mg anode, in DMF as solvent, producing CO at current yields of up to 99%, at a bias of -2.2 V vs. SCE. The total reaction was represented by[159]

$$2CO_2 + Mg \rightarrow CO + MgCO_3 \tag{39}$$

Electrocarboxylation

In carboxylation reactions involving carbon dioxide, the use of sacrificial anodes of metals such as Mg, Zn, and Al provides advantages beyond the avoidance of unwanted anodic oxidation of the carbon dioxide reduction products. The Mg^{2+}, Zn^{2+}, or Al^{3+} ions thus introduced into the electrolyte generally have beneficial effects on the yields and selectivities of the reduction products, and facilitate the isolation procedures.[156]

The electrocarboxylation of olefins has been studied in detail. Ethylene was carboxylated in DMF-tetrabutyl ammonium bromide solution, with both ethylene and carbon dioxide under pressure, using sacrificial Al anodes and stainless stell (AISI 316) cathodes. The products were dicarboxylates and monocarboxylates with the general formulae

$$^-OOC(C_2H_4)_nCOO^- \text{ and } H(C_2H_4)_nCOO^- \tag{40}$$

with n = 0, 1, 2, 3, 4.[154] With an initial ethylene/carbon dioxide molar ratio of 7 in the gas phase, at an overall pressure of about 40 atm, the C_2, C_4, C_6, and C_8 dicarboxylic acids were obtained in appreciable amounts. Under these conditions the reaction appeared to be a radical-induced ethylene telomerization, with $\cdot CO_2^-$ radical anions acting as chain initiators and terminators.[160] In solvents of low proton availability, olefins were reduced at much more positive potentials than CO_2. The main product was the β-carboxylated compound. The predominant mechanism seems to be such that CO_2 acts as an electrophile, adding to the olefin anion radical.[161]

Styrene in aprotic media under strictly anhydrous conditions underwent dicarboxylation to phenyl succinic acid with yields of around 85%, as well as producing small amounts of 3-phenyl proprionic acid.[147,162] The electrocarboxylation of acenaphthylene in anhydrous media using sacrificial Al anodes with a "gas-lift" electrochemical cell resulted in *trans*-acenaphthene-1,2-dicarboxylic acid,

Acenaphthylene   Acenaphthene-1,2-dicarboxylic acid

in 81% yield, at a current density of 40 mA cm^{-2}, with more than 80% current efficiency.[163]

The electrocarboxylation of aldehydes and ketones is of considerable practical interest, as it leads to some α-arylpropionic acids, which are used as nonsteroidal antiinflammatory agents (NSAI).[164] These 2-aryl propionic acids act by cyclooxygenase inhibition, stopping the arachidonic acid cascade to prostaglandins and thromboxane A_2, which are responsible for the inflammation mechanism.[165] Using sacrificial anodes, 6-methoxyacetonaphthone was carboxylated to 2-hydroxy-2-(6-methoxy-naphthyl)-propionic acid, which was then hydrogenated to 2-(6-methoxynaphthyl)-propionic acid, which is a most active NSAI compound.[156,166,167]

A general procedure for the electrocarboxylation of both terminal and internal alkynes with carbon dioxide was developed, using the octahedral complex of nickel(II), $Ni(bpy)_3(BF_4)_2$ (bpy = 2,2'-bipyridine) as catalyst, in an electrolyte of tetrabutyl-ammonium tetrafluoroborate in dimethylformamide solution.[168] In this reaction, alkynes were converted to α,β-unsaturated acids. When carried out in one-compartment cells, with sacrificial magnesium anodes, the process was catalytic with respect to the nickel complex. In a preparative procedure, the carboxylation of 4-octyne carried out at a constant current of 50 mA, a cathode potential of about − 1.2 V (SCE), with Mg/carbon-fiber electrodes, at atmospheric pressure of carbon dioxide, yielded (E)-2-propyl-2-hexenoic acid in 80% yield,

$$\text{n–C3H7–C}\equiv\text{C–n–C3H7} \xrightarrow{\text{+ CO2}}$$

$$\text{n–C3H7–CH=C(n–C3H7)COOH}$$

The magnesium ions released by the oxidation of the magnesium anode catalyzed the cleavage and recycling of an intermediate nickelacycle, thus enabling the $Ni(bpy)_3$ complex to react catalytically. This method provides for the conversion of alkyne hydrocarbons to substituted acrylic acids under mild conditions with high regioselectivity.[169]

1,3-Enynes, hydrocarbons containing in conjugation a double and a triple bond, were carboxylated in an electrochemical reaction performed in a single compartment cell, using a consumable Mg anode, and a carbon fiber cathode, at room temperature, in the presence of a catalytic amount of a Ni(II) complex, such as $NiBr_2$/1,2-dimethoxy-ethane + pentamethyldiethylene triamine, at 1 to 5 atm of CO_2 in DMF solution. The reaction consists of a hydrocarboxylation of the triple bond, through stereoselective *cis*-addition, leading to diene carboxylic acids,

where the R substituents represent H atoms or various alkyl groups.[170,171]

Semiquinones and quinone dianions in acetonitrile solutions, obtained by electroreduction from the quinones, react with CO_2, forming addition products. Thus, 9,10-phenanthrene-quinone in CH_3CN/[n-Bu_4N]BF_4 when electroreduced added two molecules of CO_2 to produce the bis(carbonate)$^{2-}$,[172]

The reaction was found to be chemically reversible, and may therefore in principle be a basis for the reversible abstraction of carbon dioxide from the atmosphere.

Cyclic voltammetry measurements indicated reductive coupling of CO_2 to the semiquinone anions,

$$Q + e^- = \cdot Q^- \tag{41}$$

$$\cdot Q^- + CO_2 = \cdot QCO_2^- \tag{42}$$

$$\cdot QCO_2^- + e^- = Q(CO_2)^{2-} \tag{43}$$

$$Q(CO_2)^{2-} + CO_2 = Q(CO_2)_2^{2-} \tag{44}$$

The scope of the reaction was extended to a wide range of quinones.[173]

Mediation by Macrocycles and Metal Complexes

Very high selectivity for CO formation was attained with the rhenium(I) complex Re(bpy)(CO)$_3$Cl(bpy = 2,2'-bipyridine) as mediator in DMF–H_2O (9:1) solution, on glassy carbon electrodes at a potential of -1.2 V (vs. NHE), producing carbon monoxide with 98% current efficiency.[174,175]

With the rhenium complex *fac*-Re(bpy)(CO)$_3$Cl in acetonitrile solution, the electrocatalytic reduction of CO_2 on a platinum electrode at -1.5 to -1.55 V vs. SCE resulted in the production of only CO (current efficiency about 98%) and CO_3^{2-}, in agreement with the equation,

$$2CO_2 + 2e^- \rightarrow CO + Co_3^{2-} \tag{45}$$

The cyclic voltammetry study suggested the occurrence of two electrocatalytic pathways for the reduction of CO_2 to CO, involving either initial one-electron or two-electron reduction of the above rhenium complex.[176]

The mechanism of the electrochemical reduction of CO_2 to CO at glassy carbon electrodes in acetonitrile solutions containing NEt$_4$BF$_4$ and *fac*-[Re(dmbipy)(CO)$_3$Cl], where dmbipy = 4,4'-dimethyl-2,2'-bipyridyl, was studied by an *in situ* infrared analysis. In the nominal absence of water, the one-electron reduction of the complex followed by CO_2 addition led to the formation of [Re(dmbipy)(CO)$_3$(CO$_2$H)]. In the presence of added water (10%), this complex was protonated to [Re(dmbipy)(CO)$_3$(CO$_2$H$_2$)]$^+$, which reacted with acetonitrile to form [Re(dmbipy)(CO)$_3$(MeCN)]$^+$ and CO. In these acetonitrile-water solutions (with NBu$_4$Cl as electrolyte), the rate of CO production was much higher than in the above anhydrous acetonitrile.[177]

Controlled potential electrolysis over Hg pool electrodes of solutions of [Ru(bpy)$_2$(CO)$_2$]$^{2+}$ or [Ru(bpy)$_2$(CO)Cl]$^+$ in CO_2-saturated H_2O (pH 6.0)/DMF (9:1, v/v) at -1.50 V vs. SCE produced both CO and hydrogen. On the other hand, in alkaline media, H_2O (pH 9.5)/DMF (9:1, v/v), the products were HCOO$^-$ (current efficiency 34%), CO, and H_2. The mechanism proposed involved irreversible two-electron reduction to an unstable pentacoordinated Ru(O) complex

[Ru(bpy)$_2$(CO)]° from both the above Ru complexes. This intermediate coordinated CO$_2$, and then underwent protonation and dehydroxylation, finally releasing either CO or HCOO$^-$.[178,179] Much-improved selectivity for formate production was achieved by using the [Ru(bpy)$_2$(CO)$_2$]$^{2+}$ complex in acetonitrile solution containing methylamines or phenol as proton sources. In a two-compartment electrolysis cell, with a Hg pool cathode and a Pt plate anode separated by a Nafion membrane, containing the Ru complex (0.5 mM) and Bu$_4$NClO$_4$ as electrolyte, at a potential of -1.5 V vs. SCE, the current yields for the reduction of CO$_2$ to formate in the presence of 0.2 M Me$_2$NH.HCl or PhOH were 84 and 81%, respectively, at a current density of 2 to 3 mA cm^{-2}.[180]

Electrocatalysis of carbon dioxide reduction was also obtained with the complex *cis*-[Os(bpy)$_2$(CO)H][PF$_6$] in acetonitrile containing 0.1 M tetra-*n*-butyl ammonium hexafluoro phosphate, using a Pt-mesh electrode. With anhydrous CH$_3$CN, at potentials of -1.4 to -1.6 V (vs. a NaCl-saturated calomel electrode, SSCE), CO was produced with up to 90% current efficiency, as well as small amounts of formic acid. Adding water to the electrolysis medium caused an increase in the formic acid yield, which reached a maximal current efficiency of 25% with about 0.3 M H$_2$O in the medium. The net reactions may be represented by

$$3CO_2 + 2e^- + H_2O \rightarrow CO + 2HCO_3^- \qquad (46)$$

$$2CO_2 + 2e^- + H_2O \rightarrow HCO_2^- + HCO_3^- \qquad (47)$$

Presumably even the *dry* acetonitrile contained enough water for the above equations. The proposed mechanistic pathway involved the association of CO$_2$ with the highly reduced dipyridyl complex of Os by coordination sphere expansion, forming a reactive intermediate, which either dissociated to CO or in the presence of sufficient water produced formic acid.[181] Rhodium and iridium bipyridyl complexes are also active catalysts for CO$_2$ reduction. With Rh^{III}(bpy)$_2$(trifluoromethane sulfonate)$_2^{2+}$ in acetonitrile solution containing [(*n*-Bu)$_4$N](PF$_6$) as supporting electrolyte, using a carbon cloth electrode at -1.55 V (vs. SSCE), CO$_2$ reduction to HCOOH occurred with 80% current efficiency. The turnover number reached up to 12. The proton of the formic acid was presumably derived from the decomposition of the supporting electrolyte. The electron reservoir character of 2,2'-bipyridine was suggested to promote a metal-centered reduction of carbon dioxide.[182]

In the electrochemical reduction of CO$_2$ in acetonitrile containing added water, the complex *cis*[Ru(bpy)$_2$(CO)H]$^+$ caused the production of formic acid and carbon monoxide. This same complex had previously been proposed to be the active catalyst in the photochemical reduction of CO$_2$ using [Ru(bpy)$_3$]$^{2+}$ as sensitizer (see Chapter 6).[183] The electrochemical HCOOH production was shown to be due to a catalytic cycle involving (1) one-electron reduction of the above catalyst, (2) insertion of CO$_2$ into the Ru-H bond to form a once-reduced formato complex *cis*-[Ru(bpy)$_2$(CO)(OC(O)H)]°, (3) one-electron reduction of this complex, releasing HCOO$^-$ and recycling the complex to *cis*-[Ru(bpy)$_2$(CO)H]$^+$ by the reduction of H$_2$O.[184]

By linking together covalently two bipyridine complexes of transition metals, an analogy to the molecular assemblies active in natural photosynthesis may be created. The covalently linked polypyridine Re(I)/Ru(II) complex,

in CO_2/acetonitrile solution showed a catalytic wave at -1.3 V vs. SCE, while $Re(bpy)(CO)_3Cl$ under similar conditions revealed a catalytic wave only at -1.65 V vs. SCE. The only detected reduction product was CO. The pathway for CO_2 electrochemical reduction involved an intramolecular electron transfer step from Ru(I) to Re(I).[185,186]

With $[Ni(bpy)_3]^{2+}$ in acetonitrile solutions containing tetrabutyl ammonium carbonate and saturated with CO_2, electroreduction on a glassy carbon electrode produced carbon monoxide and carbonate ions. However, due to a side reaction, a large part of these products were not released. The primary product of the two-electron reduction of $[Ni(bpy)_3]^{2+}$ is a reversible reaction leading to $[Ni^\circ(bpy)_2]$. This intermediate was proposed not only to react with CO_2 to form CO and carbonate,

$$[Ni^\circ(bpy)_2] + 2CO_2 \rightarrow [Ni(bpy)_3]^{2+} + CO + CO_3^{2-} \qquad (48)$$

but also with CO to form a carbonyl complex,

$$[Ni^\circ(bpy)_2] + 2CO \rightarrow [Ni(CO)_2(bpy)] + bpy \qquad (49)$$

Highest current efficiency, up to 55% for CO production, was achieved in constant current experiments in a single-compartment electrolysis cells.[187]

In methylene chloride solutions, with Ag(II) or Pd(II) metaloporphyrins as homogeneous catalysts, and using a glassy carbon electrode at about -1.5 V (vs. Ag reference electrode) and Pt-gauze as anode, oxalic acid was obtained as the only product of CO_2 reduction, at a current density of about 3 mA cm^{-2}.[188]

The electrochemical reduction of carbon dioxide was found to be catalyzed by iron("O")porphyrins, resulting in carbon monoxide as the main product. With tetraalkylammonium salts as supporting electrolyte, in dimethylformaldehyde solution, the porphyrins were rapidly degraded, possibly by carboxylation. Much improved stability of the iron("O")porphyrin electrocatalysts was achieved by addition of anhydrous magnesium perchlorate in the medium, and by operation of the electrolysis at a low temperature, $-40°C$. The low temperature also resulted in enhanced solubility of carbon dioxide, thus increasing the catalytic efficiency. The proposed mechanism involved the insertion of one carbon dioxide molecule into the Fe coordination sphere, which was followed at room temperature by carbon-oxygen bond breakage, releasing carbon monoxide — a process which was assisted by the electrophilic Mg^{2+} ions. At low temperatures, two carbon dioxide molecules were found to be involved in the reactive intermediate complex. The catalytic process depended on the introduction of one carbon dioxide molecule into the coordination sphere of the iron porphyrin, resulting in the intermediate production of a carbene-type complex, with electron transfer from the iron atom to the carbon dioxide group.[189-191]

Cobalt(II) tetraphenylporphyrin ($Co^{II}TPP$) in tetra-*n*-butyl ammonium tetrafluoroborate dissolved in dimethyl formamide was found to mediate the electrochemical reduction of CO_2, using glassy carbon or platinum as working electrodes. At a controlled potential of -1.5 V vs. SCE, the current density was 0.1 mA cm^{-1}, and the main reduction product was formic acid, obtained with a current efficiency of about 10%.[192]

The electrochemistry of $Co^{II}TPP$ was studied by cyclic voltammetry in hexamethylphosphoramide solutions containing tetra-butylammonium tetrafluoroborate as electrolyte, using glassy carbon electrodes. Under an argon atmosphere, the cobalt complex underwent two reversible one-electron reduction waves, corresponding respectively to the transitions $Co(II) \rightarrow Co(I)$ and $Co(I) \rightarrow Co(0)$. Under a CO_2 atmosphere, the second transition was strongly changed: the reduction current was considerably enlarged, while the reoxidation peak disappeared. The explanation proposed was that the increase in the reduction current was due to the reduction of carbon dioxide according to[193]

$$Co(0)TPP + CO_2 \rightarrow Co(I)TPP + CO_2^- \qquad (50)$$

Electrochemical reduction of carbon dioxide to carbon monoxide was the main reaction pathway also when catalyzed by some palladium triphosphine complexes in dimethylformamide or acetonitrile solutions. A variety of Pd complexes of the type $[Pd(triphos)L](BF_4)_2$ (where triphos = $PhP(CH_2CH_2PPh_2)_2$ and L = CH_3CN, $P(OMe)_3$, PEt_3, $P(CH_2OH)_3$, and PPh_3) were found to be active catalysts in the electrochemical reduction of CO_2, while the analogous Ni and Pt complexes were inactive under similar conditions. In the proposed mechanism, insertion of CO_2 into a Pd–H bond was the rate-limiting step. Since the addition of the free phosphine ligand caused an inhibitory effect, decreasing the rate of CO_2 reduction, it was concluded that dissociation of a phosphine ligand from the complex was necessary for CO_2 insertion.[194] Using complexes of $[Pd(etpC)(CH_3CN)]$ $(BF_4)_2$, where etpC represents bis(dicyclohexyl phosphinoethyl) phenylphosphine, such as,

$$
\begin{array}{c}
\quad\quad \text{—PR}_2 \quad 2^+ \\
\text{RP——Pd—S} \\
\quad\quad \text{—PR}_2'
\end{array}
$$

effective reduction of carbon dioxide to carbon monoxide was achieved, with current efficiencies of up to 85% and turnover number (moles of CO produced per mole of catalyst) of up to 130.[195,196]

A landmark discovery was the observation that CO_2 can be reversibly activated by the bifunctional complexes [Co(R-salen)M], in which R-salen = substituted salen ligand, and salen = N,N'-ethylene bis(salicylidene-aminato, and M = Li, Na, K, Cs.[197]

Co(salen) R = –CH_2CH_2– ; Co(salophen) R = –o–C_6H_4–

In the presence of alkali cations, the electrocatalytic reduction of carbon dioxide to carbon monoxide was mediated by the cobalt complexes Co(salen) and Co(salophen), carried out either in anhydrous or aqueous acetonitrile, with an alkali metal perchlorate as the base electrolyte.[198,199]

In the case of Co(salen) catalysis, this reaction was enhanced by the presence of water. The reaction was performed on a Hg-pool, at -1.4 V vs. SCE, at a current density of about 2 mA cm^{-2}, yielding both CO and HCO_3^-. Replacing $NaClO_3$ by $LiClO_3$ as the base electrolyte caused a positive shift in the CO_2 reduction potential of 110 mV.[198]

With [Co^I(salophen)Li], where salophen = N,N'-o-phenylene bis (salicylidene iminato), the addition of water decreased the selectivity to CO formation, and caused increased production of hydrogen. In anhydrous acetonitrile solution, with lithium perchlorate as base electrolyte, on a mercury electrode at -1.5 V vs. SCE, the overall reduction of CO_2 resulted in the formation of CO and carbonate, which precipitated as Li_2CO_3,

$$2CO_2 + 2e^- \rightarrow CO + CO_3^{2-} \tag{51}$$

The current efficiency for CO production was 29%, with a turnover number based on the catalyst of 22. The mechanism was represented by the reduction of Co(II) to Co(I) as the primary process (with a color change in the complex from brownish red to green), followed by reversible stepwise binding of two CO_2 molecules to the Co(I) complex. The lithium ion played an essential role in the reaction, and the intermediate complex was assumed to contain a head-to-tail CO_2 dimer, $C(O)OCO_2$, which was C-bonded to cobalt and stabilized by Li^+.[199]

Films of $[Re(CO)_3(v\text{-bpy})Cl]$, where v-bpy is 4-vinyl, 4'-methyl-2,2'-bipyridine, were deposited by electropolymerization on Pt and glassy carbon electrodes. In acetonitrile containing tetra-*n*-butyl ammonium perchlorate, CO_2 was electroreduced to CO with a current efficiency of over 95% and a turnover number of about 600.[200,201]

In a search for mediators for carbon dioxide reduction to formic acid, terdentate transition metal coordination complexes were found to be more stable than those with bidentate and tetradentate ligands. With DMF or acetonitrile as solvents, Fe, Co, and Ni complexes were tested. With the cobalt complex of dapa,

dapa   v-tpy

$Co(dapa)_2(PF_6)_2$ in DMF solution, the main CO_2 reduction product was formic acid, produced with a current efficiency of more than 60%. In case the ligand contained an olefinic group, such as in Co(v-tpy) and in Ni(v-tpy), the complex was deposited on the electrode surface by electropolymerization — forming an electroactive polymeric film of the complex. Such films were effective for the electrocatalytic reduction of carbon dioxide. Thus, the reduction potential of $Co(v\text{-tpy})_2(PF_6)_2$ was only -0.9 V (vs. SSCE). On carrying out rotating-disk electrode experiments, and applying a Koutecky-Levich plot, the second-order rate constant for the electrocatalytic reduction of carbon dioxide was determined to be $k_2 = 40$ M^{-1} s^{-1}.[202]

Tetraazamacrocyclic compounds of Co and Ni dissolved in water-acetonitrile mixtures, with Hg-cathodes at about -1.5 V (vs. SCE), mediated the reduction of CO_2, producing CO, H_2 (in the ratio 1:1), and formic acid at a total current yield of up to 98%. Carbon dioxide reduction occurred at a potential which was closer to the thermodynamic value than in the absence of the mediator.[107]

Several tetraazamacrocyclic Co(II) and Ni(II) complexes were tested as electron transfer agents in the electroreduction of CO_2 on mercury electrodes, in both aqueous and organic solvents. The products determined were CO and H_2. Highest current efficiency for CO, up to 66%, was achieved with the complex [Co(II)(Me$_2$-Pyo{14}trieneN$_4$)]$^{2+}$,

using a DMF/H_2O medium (95/5 v/v) containing 0.1 *M* Et4NCl under a CO_2 atmosphere, and with the Hg electrode at a potential of -1.30 V (vs. SCE).[203]

The rhodium complex Rh(diphos)$_2$Cl[3] (diphos = 1,2-bis (diphenyl phosphino)ethane) was found to catalyze the electroreduction of CO_2. Using a mercury pool cathode, at about -1.5 V (vs. a silver wire reference), in 0.1 *M* tetraethylammonium perchlorate in acetonitrile solution, separated by a glass frit from a Pt anode, at a current density of about 8 mA cm^{-2}, the main product was the formate anion, obtained in up to 42% current efficiency, as well as small amounts of cyanoacetate. The mechanism proposed involved the reduction of the starting rhodium complex to a neutral species Rh(diphos)$_2^\circ$, which then interacted with CO_2.[204]

A trinuclear nickel complex, [Ni$_3$(η_3-CNMe)(η_3-I)(dppm)$_3$[PF$_6$] (dppm = Ph$_2$PCH$_2$PPh$_2$), was found to be an electrocatalyst for CO_2 reduction in nonaqueous solutions. This complex has a potential for one-electron reduction at $E_{1/2} = -1.09$ V vs. Ag/AgCl. Using 0.1 *M* NaPF$_6$/MeCN as medium, on a Pt-gauze electrode at the same potential of -1.09 V vs. Ag/AgCl, the reduction of CO_2 with this electrocatalyst produced CO and CO_3^{2-}. Seven turnovers were reached within 3 h, indicating the catalytic nature of the reaction. The low reduction potential may indicate the formation of an adduct with the reduced form of the trinickel complex cluster. A second CO_2 molecule may then insert into this adduct in a "head-to-tail" configuration. After a second electron transfer, the CO_2 dimer may then disproportionate to CO and CO_3^{2-}.[205]

Electrochemical reduction of CO_2 to CO, HCOOH, and traces of CH_4 in non-aqueous solvents was catalyzed by complexes of Co(II), Ni(II), Fe(II), and Cu(II) with 1,10-*o*-phenanthroline,

Cyclic voltammetry measurements carried out in dimethyl sulfoxide solution containing 0.1 M tetrabutyl ammonium perchlorate indicated that the reduction of CO_2 was dependent both upon ligand and metal reduction on the complexes.[206]

REDUCTION IN MOLTEN SALT MEDIA

In molten salt media, electrochemical reduction may be carried out at elevated temperatures, with lowered overpotentials. Advantages of reactions in molten salts include high solubilities for many inorganic compounds and fast reaction rates because of high temperatures.[207] Most of the interest until now in the electrochemistry of carbon dioxide or of carbonates has been due to their potential application in alkaline fuel cells, or for oxygen recovery from CO_2 in life-support systems.[208]

In the molten alkali carbonate eutectic, (Li_2CO_3–Na_2CO_3–K_2CO_3 at 43.5:31.5:25.0 mol%; liquidus temperature, 397°C), at 600 to 700°C, the predominant reactions are both metal deposition,

$$M^+ + e^- \rightarrow M \tag{52}$$

and the direct cathodic reduction of carbon dioxide to CO,

$$2CO_2 + 2e^- \rightarrow CO + CO_3^{2-} \tag{53}$$

Below 600°C, carbon dioxide is mainly reduced to elementary carbon,

$$CO_3^{2-} + 4\,e^- \rightarrow C + 3O^{2-} \tag{54}$$

In K_2SO_4 melts, metal deposition is the predominant reaction at all temperature ranges. Of technical importance is the cathodic deposition of refractory carbides from molten carbonates, leading to silicon, tantalum, and tungsten carbides,[209,210]

$$2CO_3^{2-} + 10e^- \rightarrow C_2^{2-} + 6O^{2-} \tag{55}$$

Above 700°C, the change in the equilibrium,

$$C + CO_2 = 2CO \tag{56}$$

leads to carbon monoxide as the major product.[211,212]

Deposition of carbon at Au or Pd cathodes in (Li–Na–K)CO_3 (molar ratio 4:3:3) melts at 550°C at potentials of −2.5 to −2.7 (vs. Ag electrode in Ag_2SO_4 melt) was observed.[213]

Most previous studies on reactions of carbonate in molten salts have been concerned with the reduction of carbonate ions in fuel cells. The main cathodic process in fuel cells in the presence of both oxygen and carbon dioxide is the reaction of carbonate with oxygen leading to peroxide,

$$O_2 + 2CO_3^{2-} \rightarrow 2O_2^{2-} + 2CO_2 \tag{57}$$

followed by conversion of the carbon dioxide with part of the peroxide to carbonate,

$$2O_2^{2-} + 2CO_2 + 2e^- \rightarrow 2CO_3^{2-} \tag{58}$$

Thus, the overall reaction is

$$O_2 + 2e^- \rightarrow O_2^{2-} \tag{59}$$

Smooth gold and NiO electrodes were used in order to determine the exchange current density of the fuel cell cathodic oxygen reduction in the $Li_2CO_3 + K_2CO_3$ eutectic at 650°C.[214,215]

The anodic reaction at noble metal electrodes is oxidation of the carbonate ion,[216]

$$CO_3^{2-} \rightarrow CO_2 + 1/2O_2 + 2e^- \tag{60}$$

Nickel electroplated on the refractory oxides Al_2O_3, $SrTiO_3$, and $LiAlO_2$ was used for the preparation of porous nickel plates, to serve as stable anodes in molten carbonate fuel cells.[217]

An interesting application of molten alkali carbonates is in life-support systems, in the recovery of oxygen from carbon dioxide, e.g., in a spacecraft or a submarine.[208,218] For this purpose the anodic and cathodic reactions are

$$CO_3^{2-} \rightarrow CO_2 + 1/2O_2 + 2e^- \tag{61}$$

$$2CO_2 + 2e^- \rightarrow CO + CO_3^{2-} \tag{62}$$

Hence, the overall reaction is oxygen recovery,

$$CO_2 \rightarrow CO + 1/2O_2 \tag{63}$$

In the $Li_2CO_3 + Na_2CO_3 + K_2CO_3$ eutectic at 680°C, at smooth gold electrodes, the cathodic process of oxygen reduction was shown to be diffusion controlled, forming peroxide, O_2^{2-}, while oxide ions O^{2-} were formed in the anodic reaction.[219,220]

Higher current densities in carbonate melts were achieved by using alkali halide-carbonate mixtures.[221] In a KCl + NaCl eutectic at 700°C saturated with carbon dioxide under pressure, the current-voltage relationship suggested the following two-stage reduction of carbon dioxide, leading to elementary carbon,[222,223]

$$CO_2 + 2e^- \rightarrow CO_2^{2-} \tag{64}$$

$$CO_2^{2-} + 2e^- \rightarrow C + 2O^{2-} \tag{65}$$

The above mechanism was further confirmed, mainly with Ni electrodes in LiCl–KCl eutectic melts at 450°C, using the potential sweep method. The two cathodic peaks observed at 0.8 and 0.5 V (vs. Li/Li$^+$) were ascribed respectively to the reduction of carbonate ion and to the intercalation of reduced lithium into the electrodeposited carbon layer. The electrodeposited films were identified by ESCA and X-ray diffraction to be amorphous carbon.[224] The two-step mechanism of carbonate reduction was proven by the appearance of two anodic peaks in the voltammogram, indicating the reoxidation of reduced intermediates. The rate-determining two-electron transfer,

$$CO_3^{2-} + 2e^- \rightarrow CO_2^{2-} + O^{2-} \tag{66}$$

was followed by the carbon forming reaction,

$$Co_2^{2-} + 2e^- \rightarrow C + 2O^{2-} \tag{67}$$

The overall reaction was thus proposed to be represented by

$$CO_3^{2-} + 4e^- \rightarrow C + 3O^{2-} \tag{68}$$

The diffusion coefficient of the carbonate ion at 450°C was estimated to be 1.66 $\times$ 10^{-5} cm^2 s^{-1}.[225]

An example of the temperature dependence of the rate of reduction of CO_2 to CO in the alkali carbonate eutectic mixture is presented in Figure 2.[226] Silver and nickel electrodes were found to be relatively resistant to alkali carbonates and alkali halides at high temperatures. Results for CO_2 reduction to CO in various combinations of electrodes in halide melts are shown in Figure 3. With stainless steel electrodes in LiCl + KCl melts, extremely high Faradaic yields of conversion of CO_2 to Co were obtained, much larger than could be accounted for by the simple 2-electron reduction process. This was accompanied by strong corrosion of the electrodes, proving the participation of additional redox reactions.[226]

While it is feasible to reduce carbon dioxide to carbon monoxide in molten salts as electrolytes, the practical application is limited by the considerable technical problems of working with molten salts. Alkali carbonates, unless very thoroughly dried (such as by passing dry HCl through the melt), are corrosive even to quartz, porcelain, and alumina vessels. Alkali halides, such as LiCl + KCl, undergo sublimation at high temperatures, and are corrosive not only to the above containers, but also to stainless steel, copper, and platinum.

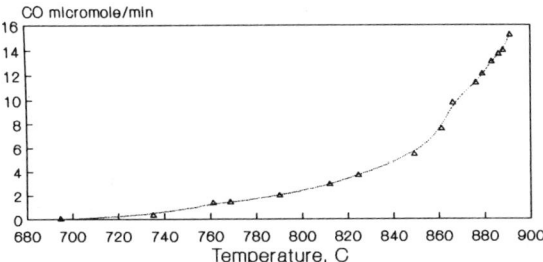

Fig. 2   Electroreduction in carbonate melt.

SOLID PHASE ELECTROLYTES

Using ceramic oxide electrolytes, such as calcium or yttrium-stabilized zirconia (YSZ), the electrolyte serves as an oxide ion conductor, effecting the cathodic release of hydrogen and the anodic release of oxygen. High-temperature proton-conductive solid electrolytes based on Y or Yb-doped $SrCeO_3$ have been used for fuel cells at 800 to 1000°C,[227] and for steam electrolysis at 800°C.[228]

The mechanism of ionic transport in calcium or YSZ at high temperatures depends on oxygen vacancies, which enable the migration of oxide ions O^{2-} through the lattice toward the anode, at which the oxide ion may be oxidized to molecular oxygen.[229] The lowest resistivities of cubic solid solutions of yttria or calcium in zirconia at about 1000°C were obtained when they contained the minimum amount of the lower-valent oxide needed to stabilize the solution.[230] Electrolysis of water at high temperatures with solid electrolytes has been investigated in numerous studies.[231,232] Using ceramic oxide electrolytes, such as calcium or YSZ, the electrolyte served as an oxide ion conductor, effecting the cathodic release of hydrogen,

$$H_2O + 2e^- \rightarrow O^{2-} + H_2 \qquad (69)$$

and the anodic release of oxygen:

$$O^{2-} \rightarrow 1/2O_2 + 2e^- \qquad (70)$$

so that the overall process is

$$H_2O \rightarrow H_2 + 1/2O_2 \qquad (71)$$

A somewhat analogous process is the methanation of carbon monoxide or carbon dioxide, using nickel electrodes and YSZ electrolyte, which was shown to proceed with high efficiency at 600 to 700°C.[233,234]

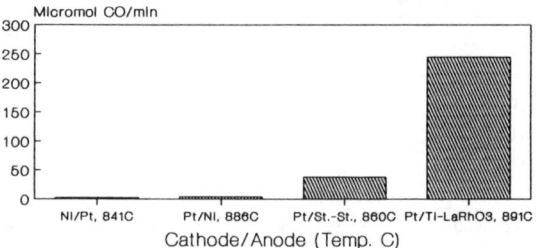

Fig. 3 Electroreduction in LiCl-KCl melt.

Direct thermal splitting of carbon dioxide at high temperatures according to

$$CO_2 = CO + 1/2O_2 \tag{72}$$

could be attractive as a means for the conversion of concentrated solar energy and its storage as a gaseous fuel. The extraction of the oxygen released from the CO and the unreacted CO_2 was possible using an oxygen semipermeable membrane, such as calcium-stabilized zirconia. The reaction cell consisted of a closed-end tube of calcium-stabilized zirconia (2 mm thick and 10-mm inner diameter). The inner part of the tube was fed with pure CO_2, at a flow rate of 100 cm^3 h^{-1}, and a pressure of 0.1 mPa. The whole assembly was in an electric oven at 1700 to 2100 K. Through an outer compartment surrounding the semipermeable cell, a CO-CO$_2$ gas mixture was pumped. Thus, there was a gradient of oxygen partial pressure from inside to outside of the semipermeable cell, which depended on the CO_2/CO ratio in the outer compartment. At the lowest ratio tested, CO_2/CO = 0.01, at 1984 K, 21.5 mol% of CO_2 was converted to CO. With a CO_2 consumption of 1000 cm^3 h^{-1}, at 1960 K, about 100 to 200 cm^3 h^{-1} of CO were produced. Even higher permeability, due to higher oxide ion conductivity, was achieved with a ceria-doped zirconia membrane.[235]

In the reverse process, oxygen gas sensors based on the measurement of the electromotive force of cells using stabilized zirconia have been applied in the steel industry and in combustion control. Similar gas sensors have been developed for the detection and assay of CO_2, using as a solid electrolyte either K_2CO_3,[236] or Na_2CO_3 coated on sodium β/β''-alumina or $Na_3Zr_2Si_2PO_{12}$, which involved cationic conductance.[237,238]

In an improved carbon dioxide sensor, a sodium ion conductor was applied, using a binary carbonate auxiliary electrode. The sensor consisted of a disk of a sodium ion conductor (NASICON, $Na_3Zr_2Si_2PO_{12}$). The CO_2-sensing surface was coated with platinum black and platinum mesh followed by a layer of a mixture of $BaCO_3$ + Na_2CO_3 (46 atom % Ba), which was fixed to the disc by melting and quenching. This sensor had a quick CO_2 response time, less than 8 s at 550°C, and was not affected by water vapor. Also, the sensor followed the Nernst equation, i.e., had excellent linear relationship between the logarithm of the CO_2 concentration, in the range of 250 to 2000 ppm CO_2, and the electromotive force. The high selectivity of this sensor for CO_2 was indicated by its being unaffected by the presence of CO (up to 1000 ppm) and of NO (up to 50 ppm). Similarly, improved CO_2 sensors were constructed using $SrCO_3$ or $CaCO_3$ instead of $BaCO_3$ in the binary carbonate. X-ray diffraction (XRD) analysis of the $BaCO_3$–Na_2CO_3 melt of the sensor indicated the absence of free Na_2CO_3 and instead the appearance of microneedle deposits. These microneedles, containing both Na and Ba atoms, were proposed to be responsible for the excellent resistance to water vapor.[239]

β-Alumina solid electrolytes operated at 450°C have been employed for the potentiometric determination of carbon dioxide partial pressures in the presence of oxygen. These sensors depend on fast sodium ion conductance. A surface layer of Na_2CO_3 on the solid electrolyte served to relate the activity of sodium to the CO_2 activity of the gas.[240] Another Na^+ conductor (NASICON) solid electrolyte CO_2 sensor used a Li-based binary carbonate auxiliary electrode containing the Li_2CO_3–$CaCO_3$ (1.8:1) eutectic mixture, operated at 500°C.[241] Lithium conductance in a solid electrolyte enabled lowering of the operating temperature of CO_2 gas sensors.[242]

The anodic reaction in CO_2–O_2 atmospheres is

$$Na_2CO_3 \rightarrow 2Na^+ + CO_2 + 1/2O_2 + 2e^- \qquad (73)$$

In CO–CO_2 atmospheres, the anodic reaction was found to be

$$Na_2CO_3 + CO \rightarrow 2Na^+ + 2CO_2 + 2e^- \qquad (74)$$

The Na^+ ions migrate through the solid electrolyte to the cathode, at which the reaction is

$$2Na^+ + 1/2O_2 + 2e^- \rightarrow Na_2O \qquad (75)$$

A different type of CO_2 sensor depends on the high sensitivity of certain mixed oxide capacitors on the CO_2 concentration. Optimal performance was reported for CuO–$BaSnO_3$ operated at 830 K.[243]

The regeneration of oxygen from carbon dioxide has been developed for closed-cycle life-support systems. Using Th, Y, or La oxide solid electrolytes at 400 to 900°C, the cathodic reaction was,[218]

$$2CO_2 + 2e^- \rightarrow CO + CO_3^{2-} \qquad (76)$$

while the anodic reaction was

$$CO_3^{2-} \rightarrow CO_2 + 1/2O_2 + 2e^- \qquad (77)$$

so that the net reaction was

$$CO_2 \rightarrow CO + 1/2O_2 \qquad (78)$$

ELECTROCHEMICAL CONCENTRATION

For recovery of carbon dioxide in low concentration from exhaust and waste gases, alkali treatment is one of the possible methods. However, for efficient reduction on most metal electrodes, the carbonate or bicarbonate must be acidified to release the electroreducible carbon dioxide. An ingenious electrode assembly to achieve this was demonstrated by anodically generated acid at a small platinum plate. Carbon dioxide was cathodically reduced at the surface of a mercury pool electrode. The anolyte (1 mol dm^{-3} H_2SO_4) was separated from the catholyte (0.8 mol dm^{-3} $NaHCO_3$ through which CO_2 was bubbled) by a glass frit. At a cathodic current density of 10 mA cm^{-2}, with magnetic stirring of the catholyte, at 0°C, the current efficiency for reduction of sodium hydrogen carbonate to formic acid was 75%.[244]

In order to recover carbon dioxide at its low concentration in the atmosphere, a process was proposed which depends on the cathodic reduction of a carrier molecule (RO) to a negatively charged state (RO^-), which then binds CO_2.

$$RO + e^- = RO^- \qquad (79)$$

$$RO^- + CO_2 = RCO_2^- \qquad (80)$$

This was then circulated to the anodic compartment, where it was oxidized to the neutral species (RCOO), which may have a lower binding constant for carbon dioxide, and thus may release the gas at a higher pressure than the initial pressure,

$$RCO_2^- = RCO_2 + e^- \qquad (81)$$

$$RCO_2 = RO + CO_2 \qquad (82)$$

Among various chemical systems tested as potential carriers of carbon dioxide, only substituted benzoquinones, such as tetrachlorobenzoquinone, had the desired properties of reversible reduction and oxidation reactions and CO_2 binding constants. However, although these quinones were able to pump CO_2 efficiently, they were too unstable for practical applications.[245,246]

References

1. **Ziessel, R.,** Chimie de coordination de la molecule de dioxyde de carbon: activation biologique, chimique, electrochimique et photochimique, *Nouv. J. Chim.,* 7, 613–633, 1983.

2. **Teeter, T. E. and Van Rysselberghe, P.,** Reduction of carbon dioxide on mercury cathodes, *Proc. 6th Meet. Int. Committee of Electrochemical Thermodynamics and Kinetics,* Poitiers, Butterworths, London, 1954, 538–542.

3. **Beketov, N. N.,** *Zh. Russ. Fiz.-Khim. Obshch.,* 1, 33, 1869.

4. **Royer, E.,** Reduction of carbonic acid into formic acid, *C. R. Acad. Sci.,* 70, 731–735, 1870.

5. **Coehn, A. and Jahn, S.,** On the electrolytic reduction of carbonic acid, *Chem. Ber.,* 37, 2836–2842, 1904.

6. **Ehrenfeld, R.,** On the electrolytic reduction of carbonic acid, *Chem. Ber.,* 38, 4138–4143, 1905.

7. **Fischer, F. and Prziza, O.,** On the electrolytic reduction of carbon dioxide and carbon monoxide dissolved under pressure, *Chem. Ber.,* 47, 256–260, 1914.

8. **Rabinowitsch, M. and Maschowetz, A.,** Electrochemical production of formate from carbonic acid, *Zeit. Elektrochem.,* 36, 846–850, 1930.

9. **Udupa, K. S., Subramanian, G. S., and Udupa, H. V. K.,** The electrolytic reduction of carbon dioxide to formic acid, *Electrochim. Acta,* 16, 1593–1598, 1971.

10. **Giner, J.,** Electrochemical reduction of CO_2 on platinum electrodes in acid solutions, *Electrochim. Acta,* 8, 857–865, 1963.

11. **Beden, B., Bewick, A., Razak, M., and Weber, J.,** On the nature of reduced CO_2. An IR spectroscopic investigation, *J. Electroanal. Chem.,* 139, 203–206, 1982.

12. **Aramata, A., Enyo, M., Koga, O., and Hori, Y.,** FT-IR spectrometry of the reduced CO_2 at Pt electrode and anomalous effect of Ca^{2+} ions, *Chem. Lett.,* 1991, 749–752.

13. **Nikolic, B. C., Huang, H., Gervasio, D., Lin, A., Fierro, C., Adzic, R. R., and Yeager, E. B.,** Electroreduction of carbon dioxide on platinum single crystal electrodes: electrochemical and insitu FTIR studies, *J. Electroanal. Chem.,* 295, 415–423, 1990.

14. **Huang, H.,** Effect of impurities on the electro-reduction of carbon dioxide on platinum electrodes in acid solutions, *J. Electrochem. Soc.,* 139, 55C–58C, 1992.

15. **Koppenol, W. H., and Rush, J. D.,** Reduction potential of the CO_2/CO_2^- couple. A comparison with other C_1 radicals, *J. Phys. Chem.,* 91, 4429–4430, 1987.

16. **Schiffrin, D. J.,** Application of photo-electrochemical effects to the study of the electrochemical properties of radicals: CO_2^- and CH_3, *Faraday Discuss. Chem. Soc.,* 56, 75–95, 1974.

17. **Kapusta, S., and Hackerman, N.,** The electroreduction of carbon dioxide and formic acid on tin and indium electrodes, *J. Electrochem. Soc.,* 130, 607–613, 1983.

18. **Aylmer-Kelly, A. W. B., Bewick, A., Cantrill, P. R., and Tuxford, A. M.,** Studies of electrochemically generated reaction intermediates using modulated specular reflectance spectroscopy, *Faraday Discuss. Chem. Soc.,* 56, 96–105, 1974.

19. **Jordan, J. and Smith, P. T.,** Free-radical intermediate in the electroreduction of carbon dioxide, *Proc. Chem. Soc.,* 1960, 246–247.

20. **Ryu, J., Andersen, T. N., and Eyring, H.,** The electrode reduction kinetics of carbon dioxide in aqueous solution, *J. Phys. Chem.,* 76, 3278–3286, 1972.

21. **Augustynski, J.,** Electrochemical reduction of carbon dioxide in aqueous solution, *Chimia,* 42, 172–175, 1988.

22. **Babenko, S. D., Benderskii, V. A., Krivenko, A. G., and Kurmaz, V. A.,** Photocurrent kinetics of the electron emission from a metal into electrolyte solution. VII. Absolute rate constants of CO_2 electrochemical reduction on mercury, *J. Electroanal. Chem.,* 159, 163–181, 1983.

23. **Vassiliev, Yu. B., Bagotzky, V. S., Osetrova, N. V., Khazova, O. A., and Mayorova, N. A.,** Electroreduction of carbon dioxide. I. The mechanism and kinetics of electroreduction of CO_2 in aqueous solutions on metals with high and moderate hydrogen overvoltages, *J. Electroanal. Chem.,* 189, 271–294, 1985.

24. **Zakharyan, A. V., Osetrova, N. V., and Vasil'ev, Yu. B.,** Adsorption of CO_2 on platinum metals, *Elektrokhimiya,* 12, 1854, 1976; *Chem. Abstr.,* 86, 129782m.

25. **Vassiliev, Yu. B., Bagotzky, V. S., Osetrova, N. V., and Mikhailova, A. A.,** Electroreduction of carbon dioxide. III. Adsorption and reduction in aprotic solvents, *J. Electroanal. Chem.,* 189, 311–324, 1985.

26. **Kaiser, U. and Heitz, E.,** On the mechanism of the electrochemical dimerization of CO_2 to oxalic acid, *Ber. Bunsenges. Phys. Chem.,* 77, 818–823, 1973.

27. **Osetrova, N. V., Vasil'ev, Yu. B., Bagotskii, V. S., Sadkova, R. G., Cherashev, A. F., and Krushch, A. P.,** Role of percarbonate in the electroreduction of carbon dioxide on platinum, *Elektrokhimiya,* 20, 286, 1984; English translation, 272.

28. **Noda, H., Ikeda, S., Yamamoto, A., Einaga, H., Yoshida, H., and Ito, K.,** Kinetics of the electroreduction of carbon dioxide on a Au cathode in phosphate buffer solution, Proc. Int. Symp. Chemical Fixation of Carbon Dioxide, Nagoya, Japan, Dec. 2–4, 1991, 327–332.

29. **Zakharyan, A. V., Rotenberg, Z. A., Osetrova, N. V., and Vasil'ev, Yu. B.,** Electroreduction of carbon dioxide on a tin electrode, *Elektrokhimiya,* 14, 1520–1527, 1978; English translation, 1317–1323.

30. **Kesarev, V. V. and Fedortsov, V. F.,** Electrochemical reduction of carbon dioxide on zinc and cadmium electrodes, *Zh. Priklad. Khim.,* 42, 707–709, 1969; English translation, 42, 673–675, 1969.

31. **Ito, K., Murata, T., and Ikeda, S.,** Electrochemical reduction of carbon dioxide to organic compounds, *Bull. Nagoya Inst. Technol.,* 27, 209–214, 1975.

32. **Hattori, A., Ikeda, S., Maeda, M., Einaga, H., and Ito, K.,** Electroreduction behavior of carbon dioxide on zinc and zinc oxide electrodes, Proc. Int. Symp. Chemical Fixation of Carbon Dioxide, Nagoya, Japan, Dec. 2–4, 1991, 323–326.

33. **Ito, K., Ikeda, S., and Okabe, M.,** Electrochemical reduction of carbon dioxide under high pressure. I. In an aqueous solution of inorganic salt, *Denki Kagaku,* 48, 247–252, 1980.

34. **Nakagawa, S., Kudo, A., Azuma, M., and Sakata, T.,** Effect of pressure on the electrochemical reduction of CO_2 on group-VIII metal electrodes, *J. Electroanal. Chem.,* 308, 339–343, 1991.

35. **Nakagawa, S., Kudo, A., Azuma, M., and Sakata, T.,** Effect of pressure on electrochemical reduction of CO_2, Proc. Int. Symp. Chemical Fixation of Carbon Dioxide, Nagoya, Japan, Dec. 2–4, 1991, 319–322.

36. **Russell, P. G., Kovac, N., Srinivasan, S., and Steinberg, M.,** The electrochemical reduction of carbon dioxide, formic acid, and formaldehyde, *J. Electrochem. Soc.,* 124, 1329–1338, 1977.

37. **Spichiger-Ulmann, M. and Augustynski, J.,** Electrochemical reduction of bicarbonate ions at a bright palladium electrode, *J. Chem. Soc., Faraday Trans. I.,* 81, 713–716, 1985.

38. **Spichiger-Ulmann, M. and Augustynski, J.,** Specific cation effect upon the cathodic reduction of bicarbonate anion at palladium, *Nouv. J. Chim.,* 10, 487–491, 1986.

39. **Spichiger-Ulmann, M. and Augustynski, J.,** Remarkable enhancement of the rate of cathodic reduction of hydrocarbonate anions at palladium in the presence of caesium cations, *Helv. Chim. Acta,* 69, 632–634, 1986.

40. **Stalder, C. J., Chao, S., and Wrighton, M. S.,** Electrochemical reduction of aqueous bicarbonate to formate with high current efficiency near the thermodynamic potential at a chemically derivatized electrode, *J. Am. Chem. Soc.,* 106, 3673–3675, 1984.

41. **André, J.-F. and Wrighton, M. S.,** Electrostatic binding of bicarbonate and formate in viologen-based redox polymers: importance in catalytic reduction of bicarbonate to formate, *Inorg. Chem.,* 24, 4288–4292, 1985.

42. **Azuma, M., Hashimoto, K., Watanabe, M., and Sakata, T.,** Electrochemical reduction of carbon dioxide to higher hydrocarbons in a $KHCO_3$ aqueous solution, *J. Electroanal. Chem.,* 294, 299–303, 1990.

43. **Bennett, E. M., Eggins, B. R., McNeill, J., and McMullan, E. A.,** Recycling carbon dioxide from fossil fuel combustion, *Anal. Proc.,* 1980, 356–359.

44. **Eggins, B. R., Brown, E. M., McNeill, E. A., and Grimshaw, J.,** Carbon dioxide fixation by electrochemical reduction in water to oxalate and glyoxylate, *Tetrahedr. Lett.,* 29, 945–948, 1988.

45. **Monnier, A., Augustynski, J., and Stalder, C.,** On the electrolytic reduction of carbon dioxide at TiO_2 and TiO_2-Ru cathodes, *J. Electroanal. Chem.,* 112, 383–385, 1980.

46. **Koudelka, M., Monnier, A., and Augustynski, J.,** Electrocatalysis of the cathodic reduction of carbon dioxide on platinized titanium dioxide film electrodes, *J. Electrochem. Soc.,* 131, 745–750, 1984.

47. **Tinnemans, A. H. A., Koster, T. P. M., Thewissen, D. H. W. M., De Kreuk, C. W., and Mackor, A.,** On the electrolytic reduction of carbon dioxide at TiO_2 and other titanates, *J. Electroanal. Chem.,* 145, 449–456, 1983.

48. **Augustynski, J.,** Comments on the paper "On the electrolytic reduction of carbon dioxide at TiO_2 and other titanates" by Tinnemans, A. H. A., Koster, T. P. M., Thewissen, D. H. W. M., De Kreuk, C. W., Mackor, A., *J. Electroanal. Chem.,* 145, 457–460, 1983.

49. **Nakabayashi, S. and Kira, A.,** An electrochemical reduction of CO_2 on conductive ceramics, Proc. Int. Symp. Chemical Fixation of Carbon Dioxide, Nagoya, Japan, Dec. 2–4, 1991, 291–294.

50. **Nakabayashi, S. and Kira, A.,** Electrochemical reduction of carbon dioxide on titanium diboride, *J. Electroanal. Chem.,* 319, 381–385, 1991.

51. **Hori, Y., Kikuchi, K., and Suzuki, S.,** Production of CO and CH_4 in electrochemical reduction of CO_2 at metal electrodes in aqueous hydrogen carbonate solution, *Chem. Lett.,* 1985, 1695–1698.

52. **Hori, Y., Kikuchi, K., Murata, A., and Suzuki, S.,** Production of methane and ethylene in electrochemical reduction of carbon dioxide at copper electrode in aqueous hydrogen carbonate solution, *Chem. Lett.,* 1986, 897–898.

53. **Hori, Y., Murata, A., Kikuchi, K., and Suzuki, S.,** Electrochemical reduction of CO_2 to CO at a gold electrode in aqueous $KHCO_3$, *J. Chem. Soc. Chem. Commun.,* 1987, 728–729.

54. **Hori, Y., Murata, A., and Takahashi, R.,** Formation of hydrocarbons in the electrochemical reduction of carbon dioxide at a copper electrode in aqueous solution, *J. Chem. Soc. Far. Trans. I,* 85, 2309–2326, 1989.

55. **Hori, Y., Murata, A., and Yoshinami, Y.,** Adsorption of CO, intermediately formed in electrochemical reduction of CO_2 at copper electrode, *J. Chem. Soc. Faraday Trans.,* 87, 125–128, 1991.

56. **Ito, K., Ikeda, S., Yamauchi, N., Iida, T., and Takagi, T.,** Electrochemical reduction products of carbon dioxide at some metallic electrodes in nonaqueous electrolytes, *Bull. Chem. Soc. Jpn.,* 58, 3027–3028, 1985.

57. **Ikeda, S., Takagi, T., and Ito, K.,** Selective formation of formic acid, oxalic acid, and carbon monoxide by electrochemical reduction of carbon dioxide, *Bull. Chem. Soc. Jpn.,* 60, 2517–2522, 1987.

58. **Noda, H., Ikeda, S., Oda, Y., and Ito, K.,** Potential dependencies of the products on electrochemical reduction of carbon dioxide at a copper electrode, *Chem. Lett.,* 1989, 289–292.

59. **Noda, H., Ikeda, S., Oda, Y., Imai, K., Maeda, M., and Ito, K.,** Electrochemical reduction of carbon dioxide at various metal electrodes in aqueous potassium hydrogen carbonate solution, *Bull. Chem. Soc. Jpn.,* 63, 2459–2462, 1990.

60. **Azuma, M., Hashimoto, K., Hiramoto, M., Watanabe, M., and Sakata, T.,** Electrochemical reduction of carbon dioxide on various metal electrodes in low-temperature aqueous $KHCO_3$ media, *J. Electroanal. Chem.,* 137, 1772–1778, 1990.

61. **Azuma, M., Hashimoto, K., Hiromoto, M., Watanabe, M., and Sakata, T.,** Carbon dioxide reduction at low temperatures on various metal electrodes, *J. Electroanal. Chem.,* 260, 441–415, 1989.

62. **Hori, Y. and Murata, A.,** Electrochemical evidence of intermediate formation of adsorbed CO in cathodic reduction of CO_2 at a Ni electrode, *Electrochim. Acta,* 35, 1777–1780, 1990.

63. **Murata, A. and Hori, Y.,** Product selectivity affected by cationic species in electrochemical reduction of CO_2 and CO at a Cu electrode, *Bull. Chem. Soc. Jpn.,* 64, 123–127, 1991.

64. **Ikeda, S., Amakusa, S., Noda, H., Saito, Y., and Ito, K.,** Photoelectrochemical and electrochemical formation of methane from carbon dioxide at copper coated electrodes, *Proc. Electrochem. Soc., Photoelectrochemistry and Electrosynthesis on Semiconducting Materials,* Vol. 88–14, 1988, 130–136.

65. **Murata, A. and Hori, Y.,** Formation of hydrocarbons in electrochemical reduction of carbon monoxide at an Fe electrode in connection with electrochemical reduction of carbon dioxide, *Denki Kagaku,* 59, 499–503, 1991; *Chem. Abstr.,* 115, 265251g.

66. **Koga, O., Nakama, K., Murata, A., and Hori, Y.,** Effects of surface state of copper electrode on the selectivity of electrochemical reduction of CO_2, *Denki Kagaku,* 57, 1137–1140, 1989; *Chem. Abstr.,* 112, 65318k.

67. **Kyriacou, G. and Anagnostopoulos, A.,** Electro-reduction of CO_2 on differently prepared copper electrodes — the influence of electrode treatment on the current efficiencies, *J. Electroanal. Chem.,* 322, 233–246, 1992.

68. **Koga, O., Murata, A., and Hori, Y.,** Effect of cadmium deposition on electroreduction of carbon dioxide by a copper electrode, *Nippon Kagaku Kaishi,* 1991, 873–878; *Chem. Abstr.,* 115, 101388f.

69. **Cook, R. L., Mac Duff, R. C., and Sammells, A. F.,** Electrochemical reduction of carbon dioxide to methane at high current densities, *J. Electrochem. Soc.,* 134, 1873–1874, 1987.

70. **Cook, R. L., Macduff, R. C., and Sammells, A. F.,** Efficient high rate CO_2 reduction to methane and ethylene at *in situ* electrodeposited copper electrode, *J. Electrochem. Soc.,* 134, 2375, 1987.

71. **Cook, R. L., MacDuff, R. C., and Sammells, A. F.**, Evidence for formaldehyde, formic acid and acetaldehyde as possible intermediates during electrochemical CO_2 reduction at copper electrodes, *J. Electrochem. Soc.*, 136, 1982–1984, 1989.

72. **Hori, Y.**, Electrochemical reduction of CO_2 at metallic electrodes, Proc. Int. Symp. Chemical Fixation of Carbon Dioxide, Nagoya, Japan, Dec. 2–4, 1991, 107–116.

73. **Nakato, Y., Yano, S., Yamaguchi, T., and Tsubomura, H.**, Reactions and mechanism of the electrochemical reduction of carbon dioxide on alloyed copper-silver electrodes, *Denki Kagaku*, 59, 491–498, 1991.

74. **Nakato, Y.**, The mechanism of electrochemical reduction of carbon dioxide at metal electrodes, Proc. Int. Symp. Chemical Fixation of Carbon Dioxide, Nagoya, Japan, Dec. 2–4, 1991, 117–122.

75. **Azuma, M., Kawasaki, Y., and Tamura, H.**, Electrochemical reduction of carbon dioxide in an aqueous $KHCO_3$ solution with small amounts of methanol, Proc. Int. Symp. Chemical Fixation of Carbon Dioxide, Nagoya, Japan, Dec. 2–4, 1991, 287–290.

76. **Cook, R. L., Mac Duff, R. C., and Sammells, A. F.**, On the electrochemical reduction of carbon dioxide at *in situ* electrodeposited copper, *J. Electrochem. Soc.*, 135, 1320–1326, 1988.

77. **Cook, R. L., MacDuff, R. C., and Sammells, A. F.**, Ambient temperature gas phase CO_2 reduction to hydrocarbons at solid polymer electrolyte cells, *J. Electrochem. Soc.*, 135, 1470–1471, 1988.

78. **Cook, R. L., MacDuff, R. C., and Sammells, A. F.**, Gas-phase CO_2 reduction to hydrocarbons at metal/solid polymer electrolyte interface, *J. Electrochem. Soc.*, 137, 187–189, 1990.

79. **Kim, J. J., Summers, D. P., and Frese, K. W., Jr.**, Reduction of CO_2 and CO to methane on Cu foil electrodes, *J. Electroanal. Chem.*, 245, 223–244, 1988.

80. **DeWulf, D. W., Jin, T., and Bard, A. J.**, Electrochemical and surface studies of carbon dioxide reduction to methane and ethylene at copper electrodes in aqueous solutions, *J. Electrochem. Soc.*, 136, 1686–1691, 1989.

81. **Santilli, D. S. and Castner, D. G.**, Mechanism of chain growth and product formation for the Fischer-Tropsch reaction over iron catalysts, *Energy Fuels*, 3, 8–15, 1989.

82. **Cook, R. L., MacDuff, R. C., and Sammells, A. F.**, Electrochemical Fischer-Tropsch reduction of carbon dioxide to hydrocarbons and alcohols, Proc. Int. Symp. Chemical Fixation of Carbon Dioxide, Nagoya, Japan, Dec. 2–4, 1991, 39–48.

83. **Frese, K. W., Jr. and, Leach, S.**, Electrochemical reduction of carbon dioxide to methane, methanol and CO on Ru electrodes, *J. Electrochem. Soc.*, 132, 259–260, 1985.

84. **Summers, D. P. and Frese, K. W., Jr.**, Electrochemical reduction of CO_2. Characterization of the formation of CH_4 at Ru electrodes in CO_2 saturated aqueous solutions, *Langmuir*, 4, 51–57, 1988.

85. **Fujishima, A.**, Electrochemical carbon dioxide reduction using solar energy, Proc. Int. Symp. Chemical Fixation of Carbon Dioxide, Nagoya, Japan, Dec. 2–4, 1991, 11–18.

86. **Wasmus, S., Cattaneo, E., and Vielstich, W.**, Reduction of carbon dioxide to methane and ethene — an on-line MS study with rotating electrodes, *Electrochim. Acta*, 35, 771–775, 1990.

87. **Canfield, D. and Frese, K. W., Jr.**, Reduction of carbon dioxide to methanol on n-GaAs and p-GaAs and p-InP. Effect of crystal face, electrolyte and current density, *J. Electrochem. Soc.*, 130, 1772–1773, 1983.

88. **Frese, K. W., Jr. and Canfield, D.**, Reduction of CO_2 on n-GaAs electrodes and selective methanol synthesis, *J. Electrochem. Soc.*, 131, 2518–2522, 1984.

89. **Sears, W. M. and Morrison, S. R.,** Carbon dioxide reduction on gallium arsenide electrodes, *J. Phys. Chem.,* 89, 3295–3298, 1985.

90. **Summers, D. P., Leach, S., and Frese, K. W., Jr.,** The electrochemical reduction of aqueous carbon dioxide to methanol at molybdenum electrodes with low overpotentials, *J. Electroanal. Chem.,* 205, 219–232, 1986.

91. **Arai, G., Harashina, T., and Yasumori, I.,** Selective electrocatalytic reduction of carbon dioxide to methanol on Ru-modified electrode, *Chem. Lett.,* 1989, 1215–1218.

92. **Okada, G., Kobayashi, K., and Kumanotani, J.,** Electrochemical reduction of carbon dioxide at metal doped TiO_2 electrodes, *Denki Kagaku,* 56, 651–652, 1988.

93. **Bandi, A.,** Electrochemical reduction of carbon dioxide on conductive metallic oxides, *J. Electrochem. Soc.,* 137, 2157–2160, 1990.

94. **Bandi, A. and Kühne, H.-M.,** Electrochemical reduction of carbon dioxide in water. Analysis of reaction mechanism on ruthenium-titanium oxide, *J. Electrochem. Soc.,* 139, 1605–1610, 1992.

95. **Schwarz, J., Maier, C. U., and Bandi, A.,** Adsorption of CO_2 on different transition metals and oxides, Proc. Int. Symp. Chemical Fixation of Carbon Dioxide, Nagoya, Japan, Dec. 2–4, 1991, 439–442.

96. **Ikeda, S. and Ito, K.,** Artificial photosynthetic systems for carbon dioxide fixation, Proc. Int. Symp. Chemical Fixation of Carbon Dioxide, Nagoya, Japan, Dec. 2–4, 1991, 23–30.

97. **Frese, K. W.,** Electrochemical reduction of CO_2 at intentionally oxidized copper electrodes, *J. Electrochem. Soc.,* 138, 3338–3344, 1991.

98. **Ogura, K. and Yoshida, I.,** Electrocatalytic reduction of carbon dioxide to methanol in the presence of 1,2-dihydroxybenzene-3,5-disulphonate ferrate (III) and ethanol, *J. Mol. Catal.,* 34, 67–72, 1986.

99. **Ogura, K. and Yoshida, I.,** Catalytic conversion of CO and CO_2 into methanol with a solar cell, *J. Mol. Catal.,* 34, 309–311, 1986.

100. **Ogura, K. and Takagi, M.,** Electrocatalytic reduction of carbon dioxide to methanol. IV. Assessment of the current-potential curves leading to reduction, *J. Electroanal. Chem.,* 206, 209–216, 1986.

101. **Ogura, K., Migita, C. T., and Imura, H.,** Catalytic reduction of CO_2 with a hydrogen fuel cell, *J. Electrochem. Soc.,* 137, 1730–1732, 1990.

102. **Ogura, K., Migita, C. T., and Wadaka, K.,** Homogeneous catalysis in the mediated electrochemical reduction of carbon dioxide, *J. Mol. Catal.,* 67, 161–173, 1991.

103. **Watanabe, M., Shibata, M., Kato, A., Azuma, M., and Sakata, T.,** Design of alloy catalysts for CO_2 reduction. III. The selective and reversible reduction of CO_2 on Cu alloy electrodes, *J. Electrochem. Soc.,* 138, 3382–3389, 1991.

104. **Watanabe, M., Shibata, M., and Katoh, A.,** Design of alloy catalysts for energy-efficient and selective reduction of CO_2, Proc. Int. Symp. Chemical Fixation of Carbon Dioxide, Nagoya, Japan, Dec. 2–4, 1991, 123–128.

105. **Watanabe, M., Shibata, M., Katoh, A., Sakata, T., and Azuma, M.,** Design of alloy electrocatalysts for CO_2 reduction. Improved energy efficiency, selectivity, and reaction rate for the CO_2 electroreduction on Cu alloy electrodes, *J. Electroanal. Chem.,* 305, 319–328, 1991.

106. **Watanabe, M., Shibata, M., Katoh, A., Azuma, M., and Sakata, T.,** Design of alloy electrocatalysts for CO_2 reduction. I. The selective and reversible reduction of CO_2 at Cu-Ni alloy electrodes, *Denki Kagaku,* 59, 508–516, 1991; *Chem. Abstr.,* 115, 265252h.

107. **Fisher, B. J. and Eisenberg, R.,** Electrocatalytic reduction of carbon dioxide by using macrocycles of nickel and cobalt, *J. Am. Chem. Soc.,* 102, 7361–7363, 1980.

108. **Takahashi, K., Hiratzuka, K., Sasaki, H., and Toshima, S.,** Electrocatalytic behavior of metal porphyrins in the reduction of carbon dioxide, *Chem. Lett.,* 1979, 305–308.

109. **Cao, X., Mu, Y., Wang, M., and Luan, L.,** The electrocatalytic reduction of carbon dioxide using cobalt tetrakis (4-trimethyl ammonio phenyl)-porphyrin under high pressure, *Huaxue Xuebao,* 1986, 44, 220–224, 1986; *Chem. Abstr.,* 104, 195348r.

110. **Cao, X., Zheng, G., and Teng, Y.,** Electrocatalytic reduction of carbon dioxide. IV. Studies on the mechanism of the electrocatalytic reaction by optically transparent thin layer electrode (OTTLE), *Huaxue Xuebao,* 47, 575–582, 1989; *Chem. Abstr.,* 111, 122682e.

111. **Beley, M., Collin, J.-P., Ruppert, R., and Sauvage, J.-P.,** Nickel (II)-cyclam: an extremely selective electrocatalyst for reduction of CO_2 in water, *J. Chem. Soc. Chem. Commun.,* 1984, 1315–1316.

112. **Beley, M., Collin, J.-P., Ruppert, R., and Sauvage, J. P.,** Electrocatalytic reduction of CO_2 by Ni-cyclam^{2+} in water. Study of the factors affecting the efficiency and the selectivity of the process, *J. Am. Chem. Soc.,* 108, 7461–7467, 1986.

113. **Collin, J.-P., Jouaiti, A., and Sauvage, J.-P.,** Electrocatalytic properties of Ni(cyclam)$^{2+}$ and Ni$_2$(biscyclam)$^{4+}$ with respect to CO_2 and H_2O reduction, *Inorg. Chem.,* 27, 1990–1993, 1988.

114. **Taniguchi, I.,** Electrocatalytic reduction of greenhouse gases using biofunctional metal complexes, Proc. Int. Symp. Chemical Fixation of Carbon Dioxide, Nagoya, Japan, Dec. 2–4, 1991, 81–88.

115. **Balazs, G. B. and Anson, F. C.,** The adsorption of Ni(cyclam)$^+$ at mercury electrodes and its relation to the electrocatalytic reduction of CO_2, *J. Electroanal. Chem.,* 322, 325–345, 1992.

116. **Hirata, Y., Suga, K., and Fujihara, M.,** Electrocatalytic reduction of CO_2 on modified electrodes with alkylcyclam-metal complex Langmuir-Blodgett films, *Thin Solid Films,* 179, 95–101, 1989.

117. **Akiba, U., Nakamura, Y., Suga, K., and Fujihara, M.,** Electrocatalytic reduction of CO_2 on Langmuir-Blodgett film modified electrode with Ni(II) complexes of amphiphilic cyclam, Proc. Int. Symp. Chemical Fixation of Carbon Dioxide, Nagoya, Japan, Dec. 2–4, 1991, 339–342.

118. **Akiba, U., Nakamura, Y., Suga, K., and Fujihira, M.,** Electrocatalytic reduction of CO_2 on a modified electrode with Langmuir-Blodgett films of nickel(II) complex with long chain alkyl substituted cyclam, *Thin Solid Films,* 210, 381–383, 1992.

119. **Hirata, Y., Suga, K., and Fujihira, M.,** *In-situ* analysis of products in electrocatalytic reduction of CO_2 with Ni-cyclam by differential electrochemical mass spectrometry during cyclic voltammetry on an amalgamated-gold mesh electrode, *Chem. Lett.,* 1990, 1155–1158.

120. **Fujihara, M., Nakamura, Y., Hirata, Y., Akiba, U., and Suga, K.,** Electrocatalytic reduction of carbon dioxide by nickel (II) complexes of cyclam and C-alkylated and N-alkylated cyclams, *Denki Kagaku,* 59, 532–539, 1991.

121. **Fujihira, M. and Noguchi, T.,** *In situ* analysis of electrochemical reduction products of CO_2 by DEMS, Proc. Int. Symp. Chemical Fixation of Carbon Dioxide, Nagoya, Japan, Dec. 2–4, 1991, 97–102.

122. **Lieber, C. M. and Lewis, N. S.,** Catalytic reduction of CO_2 at carbon electrodes modified with cobalt phthalocyanine, *J. Am. Chem. Soc.,* 106, 5033–5034, 1984.

123. **Kapusta, S. and Hackerman, N.,** Carbon dioxide reduction at a metal phthalocyanine catalyzed carbon electrode, *J. Electrochem. Soc.,* 131, 1511–1514, 1984.

124. **Meshitsuka, S., Ichikawa, M., and Tamaru, K.,** Electrocatalysis by metal phthalocyanine in the reduction of carbon dioxide, *J. Chem. Soc. Chem. Commun.,* 1974, 158–159.

125. **Rollmann, L. D. and Iwamoto, R. T.,** Electrochemistry electron paramagnetic resonance, and visible spectra of cobalt, nickel, copper, and metal-free phthalocyanines in dimethyl sulfoxide, *J. Am. Chem. Soc.,* 90, 1455–1463, 1968.

126. **Taube, R.,** The electronic structure of anionic phthalocyanine complexes of some 3d elements, *Z. Chem.,* 6, 8–21, 1966.

127. **Taube, R.,** New aspects of the chemistry of transition metal phthalocyanines, *Pure Appl. Chem.,* 38, 427–438, 1974.

128. **Furuya, N. and Koide, S.,** Electroreduction of carbon dioxide by metal phthalocyanines, *Electrochim. Acta,* 36, 1309–1313, 1991.

129. **Tanabe, H. and Ohno, K.,** Electrocatalysis of metal phthalocyanine thin film prepared by the plasma-assisted deposition on a glassy carbon in the reduction of carbon dioxide, *Electrochim. Acta,* 32, 1121–1124, 1987.

130. **Atoguchi, T., Aramata, A., Kazusaka, A., and Enyo, M.,** Cobalt(II)-tetraphenyl porphyrin-pyridine complex fixed on a glassy carbon electrode and its prominent catalytic activity for reduction of carbon dioxide, *J. Chem. Soc. Chem. Commun.,* 1991, 156–157.

131. **Atoguchi, T., Aramata, A., Kazusaka, A., and Enyo, M.,** Electrocatalytic activity of $Co^{II}TPP$-pyridine complex modified carbon electrode for CO_2 reduction, *J. Electroanal. Chem.,* 318, 309–320, 1991.

132. **Cosnier, S., Deronzier, A., and Moulet, J.-C.,** Electrochemical coating of a platinum electrode by a poly(pyrrole) film containing the *fac*-(2,2′-bipyridine) tricarbonyl chlororhenium system. Application to electrocatalytic reduction of carbon dioxide, *J. Electroanal. Chem.,* 207, 315–321, 1986.

133. **Cosnier, S., Deronzier, A., and Moulet, J.-C.,** Electrocatalytic reduction of CO_2 on electrodes modified by (2,2′-bipyridine)-$(CO)_3Cl$ complexes bonded to polypyrrole films, *J. Mol. Catal.,* 45, 381–391, 1988.

134. **Cosnier, S., Deronzier, A., and Moulet, J.-C.,** Substitution effects on the electrochemical behaviour of the (2,2′-bipyridine) tricarbonyl chlororhenium (I) complex in solution or in polymeric form and their relation to the catalytic reduction of carbon dioxide, *New J. Chem.,* 14, 831–839, 1990.

135. **Kaneko, M., Lin, R.-J., and Yoshida, T.,** Electrocatalytic reduction of CO_2 by metal complexes incorporated into coated polymer membranes, Proc. Int. Symp. Chemical Fixation of Carbon Dioxide, Nagoya, Japan, Dec. 2–4, 1991, 103–106.

136. **Petrova, G. N. and Efimov, O. N.,** Electrocatalytic reduction of CO_2 to C_1-C_3 hydrocarbons, *Elektrokhimya,* 19, 978, 1983; English translation, 875.

137. **Savinova, E. R., Yashnik, S. A., Savinov, E. N., and Parmon, V. N.,** Gas-phase electrocatalytic reduction of CO_2 to CO on carbon gas-diffusion electrode promoted by cobalt phthalocyanine, *React. Kinet. Catal. Lett.,* 46, 249–254, 1992.

138. **Mahmood, M. N., Masheder, D., and Harty, C. J.,** Use of gas-diffusion electrodes for high-rate electrochemical reduction of carbon dioxide. I. Reduction at lead, indium- and tin-impregnated electrodes, *J. Appl. Electrochem.,* 17, 1159–1170, 1987.

139. **Dewulf, D. W. and Bard, A. J.,** The electrochemical reduction of carbon dioxide to methane and ethene at copper/Nafion electrodes (solid polymer electrolyte structures), *Catal. Lett.,* 1, 73–79, 1988.

140. **Ito, T., Ikeda, S., Maeda, M., Yosida, H., and Ito, K.,** Electrochemical reduction of carbon dioxide on Cu-loaded gas diffusion electrodes, Proc. Int. Symp. Chemical Fixation of Carbon Dioxide, Nagoya, Japan, Dec. 2–4, 1991, 313–318.

141. **Mahmood, M. N., Masheder, D., and Harty, C. J.,** Use of gas-diffusion electrodes for high-rate electrochemical reduction of carbon dioxide. II. Reduction at metal phthalocyanine-impregnated electrodes, *J. Appl. Electrochem.,* 17, 1223–1227, 1987.

142. **Furuya, N. and Matsui, K.,** Electroreduction of carbon dioxide on gas-diffusion electrodes modified by metal phthalocyanines, *J. Electroanal. Chem.,* 271, 181–191, 1989.

143. **Masheder, D. and Williams, K. P. J.,** Raman spectro-electrochemistry. I. *In situ* Raman studies of the electrochemical reduction of CO_2 at lead-impregnated PTFE-bonded carbon gas diffusion electrode, *J. Raman Spectrosc.,* 18, 387–390, 1987.

144. **Gressin, J. C., Michelet, D., Nadjo, L., and Savéant, J. M.,** Electrochemical reduction of carbon dioxide in weakly protic medium, *Nouv. J. Chim.,* 3, 545–554, 1979.

145. **Amatore, C. and Savéant, J.-M.,** Mechanism and kinetic characteristics of the electrochemical reduction of carbon dioxide in media of low proton availability, *J. Am. Chem. Soc.,* 103, 5021–5023, 1981.

146. **Amatore, C., Nadjo, L., and Savéant, J. M.,** A propos de la reduction electrochimique du dioxide de carbone, *Nouv. J. Chim.,* 8, 565–566, 1984.

147. **Lamy, E., Nadjo, L., and Savéant, J. M.,** Standard potential and kinetic parameters of the electrochemical reduction of carbon dioxide in dimethylformamide, *J. Electroanal. Chem.,* 78, 403–407, 1977.

148. **Eggins, B. R. and McNeill, J.,** Voltammetry of carbon dioxide. I. A general survey of voltammetry at different electrode materials in different solvents, *J. Electroanal. Chem.,* 148, 17–24, 1983.

149. **Vassiliev, Yu. B., Bagotzky, V. S., Khazova, O. A., and Mayorova, N. A.,** Electroreduction of carbon dioxide. II. The mechanism of reduction in aprotic solvents, *J. Electroanal. Chem.,* 189, 295–309, 1985.

150. **Mayorova, N. A., Khazova, O. A., and Vassiliev, Yu. B.,** Electroreduction of carbon dioxide in aprotic solvents, *Elektrokhimiya,* 22, 1196–1204, 1986; English translation, 1122–1129.

151. **Desilvestro, J. and Pons, S.,** The cathodic reduction of carbon dioxide in acetonitrile. An electrochemical and infrared spectroelectrochemical study, *J. Electroanal. Chem.,* 267, 207–220, 1989.

152. **Ito, K., Ikeda, S., Iida, T., and Nomura, A.,** Electrochemical reduction of carbon dioxide dissolved under high pressure. III. In nonaqueous electrolytes, *Denki Kagaku,* 50, 463–469, 1982.

153. **Ito, K., Ikeda, S., Iida, T., and Niwa, H.,** Electrochemical reduction of carbon dioxide dissolved under pressure. II. In aqueous solutions of tetraalkylammonium salts, *Denki Kagaku,* 49, 106–112, 1981.

154. **Gambino, S. and Silvestri, G.,** On the electrochemical reduction of carbon dioxide and ethylene, *Tetrahedron Lett.,* 32, 3025–3028, 1973.

155. **Silvestri, G.,** Electrochemical synthesis of carboxylic acids from carbon dioxide, *NATO ASI Ser. C,* 1987, 206, 339–369, 1981.

156. **Silvestri, G., Gambino, S., and Filardo, G.,** Use of sacrificial anodes in synthetic electrochemistry. Processes involving carbon dioxide, *Acta Chem. Scand.,* 45, 987–992, 1991.

157. **Fischer, J., Lehmann, Th., and Heitz, E.,** The production of oxalic acid from CO_2 and H_2O, *J. Appl. Electrochem.,* 11, 743–750, 1981.

158. **Goodridge, F. and Presland, G.,** The electrolytic reduction of carbon dioxide and monoxide for the production of carboxylic acids, *J. Appl. Electrochem.,* 14, 791–796, 1984.

159. **Massebieau, M.-C., Duñach, E., Troupel, M., and Perichon, J.,** Efficient electrochemical reduction of carbon dioxide on freshly coated metal electrodes, *New J. Chem.,* 14, 259–260, 1990.

160. **Silvestri, G., Gambino, S., and Filardo, G.**, Electrocarboxylation of ethylene. Synthesis of industrially significant dicarboxylic acids, in *The Electrochemical Society Meeting*, Washington, DC, Abstract No. 278, 1976, 698–699.

161. **Lamy, E., Nadjo, L., and Savéant, J. M.**, On the electrochemical carboxylation of activated olefins, *Nouv. J. Chim.*, 3, 21–29, 1979.

162. **Gambino, S., Gennaro, A., Filardo, G., Silvestri, G., and Vianello, E.**, Electrochemical carboxylation of styrene, *J. Electrochem. Soc.*, 134, 2172–2175, 1987.

163. **Gambino, S., Filardo, G., and Silvestri, G.**, Electrochemical carboxylation of organic substrates. Synthesis of carboxylic derivatives of acenaphthalene, *J. Appl. Electrochem.*, 12, 549–555, 1982.

164. **Ikeda, Y. and Manda, E.**, Synthesis of benzilic acids through electrochemical reductive carboxylation of benzophenones in the presence of carbon dioxide, *Bull. Chem. Soc. Jpn.*, 58, 1723–1726, 1985.

165. **Rieu, J. P., Boucherle, A., Cousse, H., and Mouzin, G.**, Methods for the synthesis of antiinflammatory 2-aryl propionic acids, *Tetrahedron*, 42, 4095–4131, 1986.

166. **Silvestri, G., Gambino, S., and Filardo, G.**, U.S. Pat., No. 4,708,780, 1987.

167. **Maspero, F., Piccolo, O., Romano, U., and Gambino, S.**, Eur. Pat. Appl., No. 286,944, 1988.

168. **Dérien, S., Duñach, E., and Périchon, J.**, From stoichiometry to catalysis: electroreductive coupling of alkynes and carbon dioxide with nickel-bipyridine complexes. Magnesium ions as the key for catalysis, *J. Am. Chem. Soc.*, 113, 8447–8454, 1991.

169. **Dérien, S., Clinet, J.-C., Duñach, E., and Périchon, J.**, First example of direct carbon dioxide incorporation into 1,3-diynes: a highly regio selective and stereo-selective nickel catalysed electrochemical reaction, *J. Chem. Soc. Chem. Commun.*, 1991, 549–550.

170. **Dérien, S., Clinet, J. C., Duñach, E., and Périchon, J.**, Electrochemical incorporation of carbon dioxide into alkenes by nickel complexes, *Tetrahedron*, 48, 5235–5248, 1992.

171. **Dérien, S., Clinet, J.-C., Duñach, E., and Périchon, J.**, New C-C bond formation through the nickel-catalyzed electrochemical coupling of 1,3-enynes and carbon dioxide, *J. Organometal. Chem.*, 424, 213–224, 1992.

172. **Mizen, M. B. and Wrighton, M. S.**, Reductive addition of CO_2 to 9,10-phenanthrenequinone, *J. Electrochem. Soc.*, 136, 941–946, 1989.

173. **Nagaoka, T., Nishii, N., Fujii, K., and Ogura, K.**, Mechanisms of reductive addition of CO_2 to quinones, *J. Electroanal. Chem.*, 322, 383–389, 1992.

174. **Hawecker, J., Lehn, J.-M., and Ziessel, R.**, Electrocatalytic reduction of carbon dioxide mediated by Re(bipy)(CO)$_3$Cl (bipy = 2,2′-bipyridine), *J. Chem. Soc. Chem. Commun.*, 1984, 328–330.

175. **Hawecker, J., Lehn, J.-M., and Ziessel, R.**, Photochemical and electrochemical reduction of carbon dioxide to carbon monoxide mediated by (2,2′-bipyridine) tricarbonyl chlororhenium (I) and related complexes as homogeneous catalysts, *Helv. Chim. Acta*, 69, 1990–2012, 1986.

176. **Sullivan, B. P., Bolinger, C. M., Conrad, D., Vining, W. J., and Meyer, T. J.**, One- and two-electron pathways in the electrocatalytic reduction of CO_2 by fac-Re (bpy) (CO)$_3$Cl (bpy = 2,2′-bipyridine), *J. Chem. Soc. Chem. Commun.*, 1985, 1414–1416.

177. **Christensen, P., Hamnett, A., Muir, A. V. G., and Timney, J. A.**, An *in situ* infrared study of CO_2 reduction catalyzed by rhenium tricarbonyl bipyridyl derivatives, *J. Chem. Soc. Dalton Trans.*, 1992, 1455–1463.

178. **Ishida, H., Tanaka, K., and Tanaka, T.**, The electrochemical reduction of CO_2 catalyzed by ruthenium carbonyl complexes, *Chem. Lett.*, 1985, 405–406.

179. **Ishida, H., Tanaka, K., and Tanaka, T.**, Electrochemical CO_2 reduction catalyzed by [Ru(bpy)$_2$(CO)$_2$(CO)$_2$]$^{2+}$ and [Ru(bpy)$_2$(CO)Cl]$^+$. The effect on the formation of CO and $HCOO^-$, *Organometallics*, 6, 181–186, 1987.

180. **Ishida, I., Tanaka, H., Tanaka, K., and Tanaka, T.,** Selective formation of HCOO$^-$ in the electrochemical CO_2 reduction catalysed by $[Ru(bpy)_2(CO)_2]^{2+}$, *J. Chem. Soc. Chem. Commun.*, 1987, 131–132.

181. **Bruce, M. R. M., Megehee, E., Sullivan, B. P., Thorp, H., O'Toole, T. R., Downard, A., and Meyer, T. J.,** Electrocatalytic reduction of CO_2 by associative activation, *Organometallics*, 7, 238–240, 1988.

182. **Bolinger, C. M., Story, N., Sullivan, B. P., and Meyer, T. J.,** Electrocatalytic reduction of carbon dioxide by 2,2'-bipyridine complexes of rhodium and iridium, *Inorganic Chemistry*, 27, 4582–4587, 1988.

183. **Hawecker, J., Lehn, J. M., and Ziessel, R.,** Photochemical reduction of CO_2 to formate mediated by ruthenium bipyridine complexes as homogeneous catalysts, *J. Chem. Soc. Chem. Commun.*, 1985, 56–58.

184. **Pugh, J. R., Bruce, M. R. M., Sullivan, B. P., and Meyer, T. J.,** Formation of a metal hydride bond and the insertion of CO_2. Key steps in the electrocatalytic reduction of carbon dioxide to formate anion, *Inorg. Chem.*, 30, 86–91, 1991.

185. **Furue, M., Yoshidzumi, T., Kinoshita, S., Kushida, T., Nozakura, S., and Kamachi, M.,** Intramolecular energy transfer in covalently linked polypyridine ruthenium(II)/osmium(II) binuclear complexes, *Bull. Chem. Soc. Jpn.*, 64, 1632–1640, 1991.

186. **Furue, M., Maruyama, K., Naiki, M., and Kamachi, M.,** CO_2 reduction by covalently-linked binuclear complexes comprising polypyridine Re(I)/Ru(II) complexes, Proc. Int. Symp. Chemical Fixation of Carbon Dioxide, Nagoya, Japan, Dec. 2–4, 1991, 435–436.

187. **Daniele, S., Ugo, P., Bontempelli, G., and Fiorani, M.,** An electroanalytical investigation on the nickel-promoted electrochemical conversion of CO_2 to CO, *J. Electroanal. Chem.*, 219, 259–271, 1987.

188. **Becker, J. Y., Vaina, B., Eger, R., and Kaufman, L.,** Electrocatalytic reduction of CO_2 to oxalate by Ag^{II} and Pd^{II} porphyrins, *J. Chem. Soc. Chem. Commun.*, 1985, 1471–1473.

189. **Hammouche, M., Lexa, D., and Savéant, J.-M.,** Catalysis of the electrochemical reduction of carbon dioxide by iron("O")porphyrins, *J. Electroanal. Chem.*, 249, 347–351, 1988.

190. **Hammouche, M., Lexa, D., Momenteau, M., and Savéant, J.-M.,** Chemical catalysis of electrochemical reactions — homogeneous catalysis of electrochemical reduction of carbon dioxide by iron ("O") porphyrins — role of the addition of magnesium cations, *J. Am. Chem. Soc.*, 113, 8455–8466, 1991.

191. **Savéant, J.-M.,** Molecular catalysis of the electrochemical reduction of carbon dioxide. Iron "O" porphyrins, Proc. Int. Symp. Chemical Fixation of Carbon Dioxide, Dec. 2–4, 1991, Nagoya, Japan, 49–54.

192. **Atoguchi, T., Aramata, A., Kazusaka, A., and Enyo, M.,** Electrochemical reduction of CO_2 mediated by CoIITPP/CoITPP redox complex on GC and Pt in DMF, *Denki Kagaku*, 59, 526–527, 1991.

193. **Tezuka, M., Iwasaki, M., and Yajima, T.,** Voltammetric study on the electrocatalytic reduction of CO_2 in aprotic media, Proc. Int. Symp. Chemical Fixation of Carbon Dioxide, Nagoya, Japan, Dec. 2–4, 1991, 303–306.

194. **DuBois, D. L. and Miedaner, A.,** Mediated electrochemical reduction of CO_2. Preparation and comparison of an isoelectronic series of complexes, *J. Am. Chem. Soc.*, 109, 113–117, 1987.

195. **DuBois, D. L., Miedaner, A., and Haltiwanger, R. C.,** Electrochemical reduction of CO_2 catalyzed by [Pd(triphosphine) (solvent)] (BF$_4$)$_2$ complexes: synthetic and mechanistic studies, *J. Am. Chem. Soc.*, 113, 8753–8764, 1991.

196. **Bernatis, P., Curtis, C. J., Herring, A., Miedaner, A., and DuBois, D. L.,** Development of homogeneous catalysts for electrochemical CO_2 concentration and reduction, Proc. Int. Symp. Chemical Fixation of Carbon Dioxide, Nagoya, Japan, Dec. 2–4, 1991, 89–96.

197. **Gambarotta, S., Arena, F., Floriani, C., and Zanazzi, P. F.,** Carbon dioxide fixation: bifunctional complexes containing acidic and basic sites working as reversible carriers, *J. Am. Chem. Soc.,* 104, 5082–5092, 1982.

198. **Pearce, D. J. and Pletcher, D.,** A study of the mechanism for the electrocatalysis of carbon dioxide reduction by nickel and cobalt square planar complexes in solution, *J. Electroanal. Chem.,* 197, 317–330, 1986.

199. **Isse, A. A., Gennaro, A., Vianello, E., and Floriani, C.,** Electrochemical reduction of carbon dioxide catalyzed by [Co^I(salophen)Li], *J. Mol. Catal.,* 70, 197–208, 1991.

200. **Cabrera, C. R. and Abruña, H. D.,** Electrocatalysis of CO_2 reduction at surface modified metallic and semiconducting electrodes, *J. Electroanal. Chem.,* 209, 101–107, 1986.

201. **O'Toole, T. R., Sullivan, B. P., Bruce, M. R.-M., Margerum, L. D., Murray, R. W., and Meyer, T. J.,** Electrocatalytic reduction of CO_2 by a complex of rhenium in thin polymeric films, *J. Electroanal. Chem.,* 259, 217–239, 1989.

202. **Arana, C., Yan, S., Potts, K. T., and Abruña, H. D.,** Electrocatalysis of CO_2 reduction with transition metal complexes incorporating terdentate coordination, Proc. Int. Symp. Chemical Fixation of Carbon Dioxide, Nagoya, Japan, Dec. 2–4, 1991, 73–80.

203. **Tinnemans, A. H. A., Koster, T. P. M., Thewissen, D. H. M. W., and Mackor, A.,** Tetraaza-macrocyclic cobalt(II) and nickel(II) complexes as electron-transfer agents in the photo (electro) chemical and electrochemical reduction of carbon dioxide, *Rec. Trav. Chim. Pays-Bas,* 103, 288–295, 1984.

204. **Slater, S. and Wagenknecht, J. H.,** Electrochemical reduction of CO_2 catalyzed by Rh(diphos)$_2$Cl, *J. Am. Chem. Soc.,* 106, 5367–5368, 1984.

205. **Ratliff, K. S., Lentz, R. E., and Kubiak, C. P.,** Carbon dioxide chemistry of the trinuclear complex [Ni$_3$(η_3-CNMe)(η_3-I)(dppm)$_3$][PF$_6$]. Electro-catalytic reduction of carbon dioxide, *Organometallics,* 11, 1986–1988, 1992.

206. **Simpson, T. C. and Durand, R. R., Jr.,** Ligand participation in the reduction of CO_2 catalyzed by complexes of 1,10-*o*-phenanthroline, *Electrochim. Acta,* 33, 581–583, 1988.

207. **Volkov, S. V.,** Chemical reactions in molten salts and their classification, *Chem. Soc. Rev.,* 19, 21–28, 1990.

208. **Weaver, J. L. and Winnick, J.,** The molten carbonate carbon dioxide concentrator: cathode performance at high CO_2 utilization, *J. Electrochem. Soc.,* 130, 20–28, 1983.

209. **Bartlett, H. E. and Johnson, K. E.,** Electrolytic reduction and Ellingham diagrams for oxyanion systems, *Can. J. Chem.,* 44, 2119–2129, 1966.

210. **Stern, K. H. and Gadomski, S. T.,** Electrodeposition of tantalum carbide coatings from molten salts, *J. Electrochem. Soc.,* 130, 300–305, 1983.

211. **Mamantov, G.,** Molten salt electrolytes in secondary batteries, in *Materials for Advanced Batteries,* Murphy, D. W., Broadhead, J., Steele, B. C. H., Eds., Plenum Press, New York, 1980, 111–122.

212. **Deanhardt, M. L., Stern, K. H., and Kende, A.,** Thermal decomposition and reduction of carbonate ion in fluoride melts, *J. Electrochem. Soc.,* 133, 1148–1152, 1986.

213. **Dubois, J. and Buvet, R.,** Electroactivity range and pO^{2-} scale determination in a medium of alkali carbonate melts, *Bull. Chem. Soc. France,* 1963, 2522–2526.

214. **Uchida, I., Nishina, T., Mugikura, Y., and Itaya, K.,** Gas electrode reactions in molten carbonate media. I. Exchange current density of oxygen reduction in (Li + K)CO$_3$ eutectic at 650°C, *J. Electroanal. Chem.,* 206, 229–239, 1986.

215. **Uchida, I., Mugikura, Y., Nishina, T., and Itaya, K.,** Gas electrode reactions in molten carbonate media. II. Oxygen reduction kinetics on conductive oxide electrodes in (Li + K)CO$_3$ eutectic at 650°C, *J. Electroanal. Chem.,* 206, 241–252, 1986.

216. **Janz, G. J. and Conte, A.,** Potentiostatic polarization studies in fused carbonates. I. The noble metals, silver and nickel, *Electrochim. Acta,* 9, 1269–1278, 1964.

217. **Iacovangelo, C. D.,** Stability of molten carbonate fuel cell nickel anodes, *J. Electrochem. Soc.,* 133, 2410–2416, 1986.

218. **Chandler, H. W.,** Design of a test model for a solid electrolyte carbon dioxide reduction system, *Sci. Tech. Aerospace Rep.,* 4, 1008, 1966; *Chem. Abstr.,* 66, 25343q.

219. **White, S. H. and Twardoch, U. M.,** The influence of gas composition on the oxygen electrode reaction in the molten Li$_2$CO$_3$-Na$_2$CO$_3$-K$_2$CO$_3$ eutectic mixture, *Electrochim. Acta,* 27, 1599–1607, 1982.

220. **White, S. H. and Twardoch, U. M.,** The behavior of water in molten salts, *J. Electrochem. Soc.,* 134, 1080–1088, 1987.

221. **Randin, J.-P.,** Carbon, in *Encyclopedia of Electrochemistry of the Elements,* Vol. 3, Bard, A. J., Ed., Marcel Dekker, New York, 1976, 1–291.

222. **Delimarskii, Yu. K., Shapoval, V. I., and Vasilenko, V. A.,** Significance of a kinetic process during the electroreduction of carbonate ions in fused KCl-NaCl, *Elektrokhimiya,* 7, 1301–1304, 1971; English translation, 1255–1258.

223. **Shapoval, V. I., Kushkhov, K. B., and Soloviev, V. V.,** Cation catalysis of a carbonate-ion electroreduction against a background of melted chlorides, *Ukrainsk. Khim. Zhurn.,* 51, 1263, 1985; *Chem. Abstr.,* 104, 118490d.

224. **Shimada, T., Yoshida, N., and Ito, Y.,** Cathodic reduction of carbonate ion in LiCl-KCl melt. I. Electrodeposition of carbon, *Denki Kagaku,* 59, 701–706, 1991.

225. **Shimada, T., Yoshida, N., and Ito, Y.,** Cathodic reduction of carbonate ion in LiCl-KCl melt. II. Mechanism of cathodic reduction, *Denki Kagaku,* 60, 194–199, 1992.

226. **Halmann, M. and Zuckerman, K.,** Electroreduction of carbon dioxide to carbon monoxide in molten LiCl + KCl, LiF + KF + NaF, Li$_2$CO$_3$ + Na$_2$CO$_3$ + K$_2$CO$_3$ and AlCl$_3$ + NaCl, *J. Electroanal. Chem.,* 235, 369–380, 1987.

227. **Iwahara, H., Uchida, H., and Tanaka, S.,** High temperature-type proton conductive solid oxide fuel cells using various fuels, *J. Appl. Electrochem.,* 16, 663–668, 1986.

228. **Iwahara, H., Uchida, H., and Yamasaki, I.,** High-temperature steam electrolysis using SrCeO$_3$-based proton conductive solid electrolyte, *Int. J. Hydrogen Energy,* 12, 73–77, 1987.

229. **Wagner, C.,** The mechanism of electric conduction in the Nernst glower, *Naturwissenschaften,* 31, 265–268, 1943.

230. **Dixon, J. M., LeGrange, L. D., Merten, U., Miller, C. F., and Porter, J. T., II,** *J. Electrochem. Soc.,* 110, 276–280, 1963.

231. **Obayashi, H. and Kudo, T.,** High temperature electrolysis/fuel cells: materials problems, in *Solid State Chemistry of Energy Conversion and Storage,* Goodenough, J. B., Whittingham, M. S., Eds., (Advances in Chemistry Series, No. 163), American Chemical Society, Washington, DC, 1977, 316–363.

232. **Barbi, G. B. and Mari, C. M.,** High temperature electrochemical reduction of the water molecule at cerium dioxide electrodes, *Solid State Ionics,* 15, 335–343, 1985.

233. **Gür, T. M. and Huggins, R. A.,** Methane synthesis by a solid-state ionic method, *Science,* 219, 967–969, 1983.

234. **Gür, T. M. and Huggins, R. A.,** Methane synthesis over transition metal electrodes in a solid state ionic cell, *J. Catal.,* 102, 443–446, 1986.

235. **Nigara, Y. and Cales, B.,** Production of carbon monoxide by direct thermal splitting of carbon dioxide at high temperature, *Bull. Chem. Soc. Jpn.,* 59, 1997–2002, 1986.

236. **Gauthier, M. and Chamberland, A.,** Solid-state detectors for the potentiometric determination of gaseous oxides. I. Measurement in air, *J. Electrochem. Soc.*, 124, 1579–1583, 1977.

237. **Maruyama, T., Sasaki, S., and Saito, Y.,** Potentiometric gas sensor for carbon dioxide using solid electrolytes, *Solid State Ionics*, 23, 107–112, 1987.

238. **Maruyama, T., Ye, X.-Y., and Saito, Y.,** Electromotive force of the $CO-CO_2-O_2$ concentration cell using Na_2CO_3 as a solid electrolyte at low oxygen partial pressures, *Solid State Ionics*, 23, 113–117, 1987.

239. **Miura, N., Yao, S., Shimizu, Y., and Yamazoe, N.,** Carbon dioxide sensor using sodium ion conductor and binary carbonate auxiliary electrode, *J. Electrochem. Soc.*, 139, 1384–1388, 1992.

240. **Liu, J. and Weppner, W.,** Potentiometric CO_2 gas sensor based on Na-β/β''-alumina solid electrolyte at 400°C, *Eur. J. Solid State Inorg. Chem.*, 28, 1151–1160, 1991.

241. **Yao, S., Hosohara, S., Shimizu, Y., Miura, N., Futata, H., and Yamazoe, N.,** Solid electrolyte CO_2 sensor using NASICON and Li-based binary carbonate electrode, *Chem. Lett.*, 1991, 2069–2072.

242. **Imanaka, N., Murata, T., Kawasato, T., and Adachi, G.,** The operating temperature lowering for CO_2 gas sensor with a lithium conducting solid electrolyte, *Chem. Lett.*, 1992, 103–106.

243. **Ishihara, T., Kometani, K., Mizuhara, Y., and Takita, Y.,** Mixed oxide capacitor of $CuO-BaSnO_3$ as a sensor for CO_2 detection over a wide range of concentration, *Chem. Lett.*, 1991, 1711–1714.

244. **Fujiwara, H., Konno, A., and Nonaka, T.,** Carbon dioxide fixation by electrolysis of aqueous hydrogen carbonate solution. Reduction to formic acid at a mercury cathode, *Chem. Lett.*, 1991, 1843–46.

245. **Bell, W. L., Miedaner, A., Smart, J. C., DuBois, D. L., and Verostko, C. E.,** Synthesis and evaluation of electroactive CO_2 carriers, SAE Techn.. Paper Ser. No. 881078, 18th Intersoc. Conf. on Environ. Systems, San Francisco, July 11–13, 1988.

EIGHT

Photoelectrochemical Reduction

Using semiconductor electrodes as photoelectrodes, some of the overpotential required for the reduction of carbon dioxide may be gained by the photopotential produced. Fujishima and Honda[1] pioneered in demonstrating the use of an illuminated single crystal n-TiO_2 semiconductor electrode for the photolysis of water. In semiconductors under illumination with light of wavelengths shorter than the band-gap, absorption of light quanta causes excitation of electrons from the valence band to the conductance band. With p-type semiconductors, electrons from the conductance band may interact with electron-deficient molecules such as CO_2 adsorbed on the semiconductor surface, causing reduction of these molecules. In most photoelectrochemical systems, the major reduction product was HCOOH. With some electrode and mediator combinations, selectivity to CO or methanol was obtained.

DIRECT REDUCTION

The energy band positions of several semiconductors (expressed in eV vs. vacuum) are compared in Figure 1 with the redox potentials (expressed in V vs. NHE) involved in the reduction of carbon dioxide.[2]

Characterization of carbon dioxide adsorbed on the electrode surface improved considerably with application of FTIR spectrometry for the solid/solution interface.[3] On a p-CdTe electrode in acetonitrile containing $0.1\ M$ tetrabutyl ammonium perchlorate, CO_2 was reduced, forming an adsorbed ion radical, which gave rise to IR absorption peaks at 2018, 2041, 2065, and 2090 cm^{-1}.[4]

The photoelectrolysis of carbon dioxide bubbled through an aqueous phosphate buffer solution (pH 6.8) in a cell containing an illuminated p-GaP photocathode

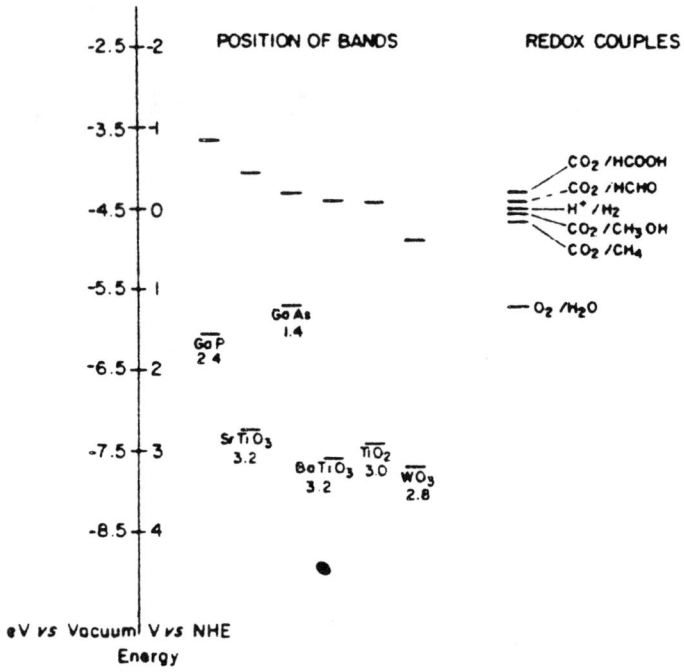

Fig. 1 (A) Semiconductor energy band positions and redox potentials and (B) photoreduction on p-GaAs; flux effect.

and a carbon counterelectrode resulted after 18 h in the production of formic acid as the main product, in 1.2×10^{-2} M concentration, as well as traces of formaldehyde and methanol. The cathode was maintained at a bias of -1.0V (vs. SCE), resulting in a current density of 1 mA cm^{-2}.[5,6] In similar experiments with either an illuminated p-GaP photocathode, or an n-TiO$_2$ electrode in the dark, at a potential of -1.5V vs. SCE, in a medium of aqueous 0.5 M H$_2$SO$_4$, the carbon dioxide reduction products were formic acid, formaldehyde, and methanol.[7] In subsequent work on p-GaP photocathodes in aqueous electrolytes, the only CO$_2$ reduction product observed was formic acid.[8]

The current efficiency for CO$_2$ reduction on p-GaP photocathodes was enhanced by electroplating or sputtering of Pb or Zn on the electrodes. Using 0.1 M aqueous Et$_4$NClO$_4$ or Et$_4$NBr as electrolytes, the only products observed were HCOOH and CO. While the current efficiency for HCOOH production was only 4.8% on the bare p-GaP electrode, it reached 48.2% on the Pb-coated p-GaP electrode, at a potential of only -1.2V vs. Ag/AgCl/saturated KCl.[9]

The light intensity effect on the reduction of CO$_2$ to HCOOH on p-GaP was studied using a xenon lamp. Maximal Faradaic efficiency of 70% was achieved at a light flux of 320 mW cm^{-2}, at an electrode potential of -0.9 V (vs. Ag/AgCl saturated KCl) in 0.1 M KHCO$_3$ saturated with CO$_2$. At constant electrode potential, the Faradaic efficiency for HCOOH production as a function of light intensity reached a maximum at about 200 to 300 mW cm^{-2}, and above that declined. The reason for this decline at high light intensity was explained by diffusion limitation for the supply of CO_2 to the electrode, and by competition with H_2 production.[10,11]

With a p-CdTe electrode in DMF containing 5% water and 0.1 M tetrabutyl ammonium perchlorate in a CO_2 atmosphere, illumination with visible light resulted in photoreduction of CO_2 to CO at potentials which were at least 0.7 V less negative than at metal electrodes. At a potential of -1.6 V vs. SCE, at a current density of about 1.8 mA cm^{-2}, the current efficiency for CO production was about 70%, while photocurrent quantum efficiency (calculated for 600 nm) was close to unity.[12] In a comparison of p-type Si, In, GaP, GaAs, and CdTe as electrodes for the reduction of carbon dioxide in DMF solutions containing 5% water and 0.1 M tetrabutyl ammonium perchlorate, best results were obtained with the (100) plane of CdTe. At a potential of -1.6 V (vs. SCE), the only products were CO and H_2. By controlled potential photoassisted electrolysis, the Faradaic efficiency for CO production was 70 to 80%, while hydrogen formation was 0.5%. Similar results were obtained using as solvents DMSO ($+5\%$ H_2O) or propylene carbonate ($+5\%$ H_2O) while much lower current yields for CO production were found with acetonitrile ($+5\%$ H_2O) as solvent.[13,14]

In a comparison of single crystal p-CdTe (111 face) and p-InP (100 face) electrodes for the photoelectrolysis of CO_2-saturated aqueous solutions, the p-InP was found to be more sensitive to corrosive deterioration than the p-CdTe. The nature of the supporting electrolyte strongly affected the product selectivity. Carbonate salts favored the production of formic acid. Other salts, such as sulfates, phosphates, and perchlorates favored the formation of CO. Highest current efficiency for CO_2 reduction was obtained at the p-CdTe electrode in aqueous solutions in the presence of tetraalkyl ammonium salts.[15]

Metal-coated p-GaP and p-InP were tested for the photoelectrochemical reduction of CO_2 in aqueous and nonaqueous electrolytes. Methane was produced by illumination of Cu-coated p-GaP electrodes. At a potential of -1.6 V (vs. Ag/AgCl), the Faradaic efficiency was 7% at 1°C.[16-18]

Electrodes of p$^+$/p-Si were compared in CO_2 and in N_2-saturated 0.5 M Na_2SO_4 solutions. The current-potential curves indicated that the rate of hydrogen generation in the CO_2-saturated solutions was larger than the rate of carbon dioxide reduction. However, when these electrodes were coated with thin films of TiO_2 rutile, grown by metalorganic chemical vapor deposition (MOCVD), the photocurrent in CO_2-saturated solutions was positively shifted by about 200 mV relative to N_2-saturated solutions. The reduction products were formic acid and formaldehyde.[19] The photoreduction of CO_2 in aqueous solutions to HCOOH on p$^+$/p-Si photocathodes was enhanced by a coating of Pb. Also, the yield and efficiency of HCOOH production depended on the pH of electrolyte. It was low in the acidic range and increased markedly above pH 6.[20]

The yields of CO_2 photoreduction products may be enhanced using elevated CO_2 gas pressures. With an electrochemical autoclave, illuminated with a 150-

W Xe-lamp through a side window of quartz, using a p-GaP photocathode at -1.0 V (vs. SSE), in 0.1 M aqueous $HClO_4$, separated by a cation exchange membrane from a platinum counterelectrode, at 3 atm pressure of CO_2, the products identified were formic acid, formaldehyde, and methanol, produced in a molar ratio of 75:6:19, and with a total Faradaic yield of 18%. In a similar experiment, but in a medium of 0.5 M $NaHCO_3$, at 7.5 atm CO_2, at a cathodic bias of -1.4 V vs. SSE, the HCOOH, HCHO, and CH_3OH product ratio was 89:3:8, with the total Faradaic yield of 38%.[21]

An interesting application of photoelectrochemistry is the electrocarboxylation of organic substrates. This has been carried out using p-GaP as a photocathode, illuminated with a 500-W xenon lamp, acetonitrile as solvent, Bu_4NBF_4 as supporting electrolyte, and phenanthrene as the reactant. During photoelectrolysis at a potential of -1.65 V (vs. SCE) under a CO_2 atmosphere, phenanthrene was carboxylated. The proposed mechanism involved prior formation of the CO_2^- radical anion, which then interacted with the aromatic compound,[22]

Selective production of methanol in the photoelectrochemical reduction of aqueous carbon dioxide was achieved using p-type copper oxide and silicon carbide.[21] The copper oxide electrode was prepared by anodic oxidation of copper metal. Under illumination (500-W xenon lamp), with a cathodic bias of -0.3 V (vs. SCE), in 0.1 M $KHCO_3$, the current efficiencies for production of methane, methanol, and formic acid were 60, 10, and 20%, respectively. After prolonged electrolysis, the current efficiency for methanol formation reached 22%. During this process, the copper oxide underwent reduction via CuO to Cu_2O and finally metallic copper, at which stage the photoreduction of carbon dioxide stopped. With single crystal p-SiC, no photoreduction products of carbon dioxide could be detected. However, using a microcrystalline SiC film electrode (which had been deposited on ITO by a mercury-assisted chemical vapor deposition method), in 0.05 M Na_2SO_4, at -1.0 V (vs. SCE), the current efficiencies for the production of H_2, CO, CH_4, and CH_3OH were 80.0, 1.6, 1.3, and 16.7%, respectively.[23]

On a Cu_2S compound semiconductor electrode (prepared by electroplating on a Cu plate from an aqueous solution of 0.1 M CuCl, 0.6 M $Na_2S_2O_3$, 1 M $NaHSO_3$ at 10 mA cm^{-2}), CH_4 was produced selectively under illumination at a potential of -0.5 V (vs. SCE), with a current efficiency of 5.3%.[24]

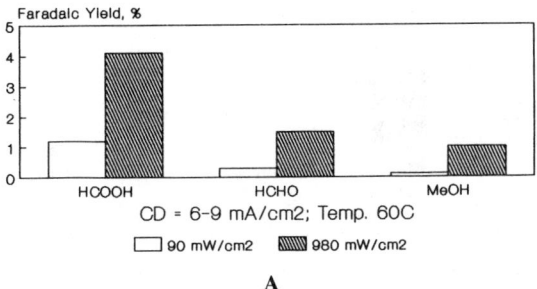

Photoreduction of CO2 on p–GaAs
Mediated by Vanadous Chloride
Effect of Light Flux.

CD = 6–9 mA/cm2; Temp. 60C

□ 90 mW/cm2 ▨ 980 mW/cm2

A

Fig. 2 Photoreduction on p-GaAs; temperature effect.

MEDIATION BY MACROCYCLES AND METAL COMPLEXES

As in the electrochemical systems, also in photoelectrochemical reactions the reduction of carbon dioxide may be enhanced by electron transfer mediators such as complexes of transition elements. Using tetraaza macrocyclic metal complexes such as $[(Me_6[14]aneN_4)Ni^{11}]^{2+}$ with a p-Si electrode in a CO_2-saturated solution of acetonitrile-H_2O-$LiClO_4$ (1:1:0.1 M), the main products were CO and H_2 formed in a 2:1 ratio, with close to 100% current efficiency. The potentials required were about 1.3 V less negative than those without the mediator.[25]

Low band-gap semiconductors, such as p-Si, p-GaAs, p-InP, and p-CdTe, are optimally matched to the solar spectrum. However, they are often unstable in aqueous media due to corrosion. This corrosion problem had been overcome with some of such low band-gap electrodes by using the vanadium redox couple V(II)-V(III), providing efficient electron transfer from cathodically polarized semiconductors. With p-GaAs photocathodes and carbon counterelectrodes, in acidic media containing vanadous chloride, carbon dioxide was photoreduced to formic acid, formaldehyde, and methanol with Faradaic efficiencies of up to 4.1, 1.5, and 1.0%, respectively. The efficiencies increased both with rising temperatures (Figure 1) and with light intensity (Figure 2).[26]

In aqueous media, using p-GaP as photocathode, Li_2CO_3 as electrolyte, in the presence of 15-crown-5, at -0.95 V vs. SCE, carbon dioxide was reported to be reduced to methanol, formic acid, and formaldehyde with current efficiencies of 44, 15, and 4%, respectively. The initial step was proposed to be electron transfer from deposited Li to CO_2, forming the CO_2^- radical anion.[27]

In an alternative approach to overcome the corrosion sensitivity of low band-gap semiconductors, polymer coatings on the semiconductor surfaces have been applied. Conductive polyaniline coatings by electropolymerization of aniline were originally developed for Pt,[28] and later for semiconductor[29] electrodes. p-Type silicon was coated with polyaniline by electropolymerization at a fixed potential

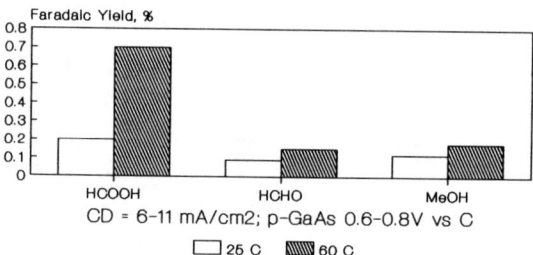

Fig. 2B (continued)

of $+0.8$ V vs. SCE from a solution of 0.1 M aniline in 0.1 M H_2SO_4, resulting in a coating of about 50 nm thickness. With this electrode, in aqueous 0.1 M $LiClO_4$ saturated with CO_2, the onset potential of the photocurrent was displaced by about $+0.3$ V relative to that of the bare p-Si electrode. The reduction products identified were formaldehyde and formic acid, in a total Faradaic yield of up to 28%, and an energy conversion efficiency (energy output/energy input) of about 4%.[30]

On the semiconductor electrodes p-Si and thin-film p-WSe$_2$, films of $[Re(CO)_3(v-bpy)Cl]$, where v-bpy is 4-vinyl-4'-methyl-2,2'-bipyridine, were deposited by electropolymerization. Carbon dioxide reduction to carbon monoxide at essentially unit current efficiency was observed upon illumination of these electrodes, in acetonitrile containing tetra-n-butyl ammonium perchlorate, with onset potentials of -0.65 V (vs. SSCE), and with turnover numbers of about 450.[31]

Selective reduction of CO_2 to CO was achieved in aqueous media by using Ni-cyclam with p-type semiconductors. Using p-GaP or p-GaAs photoelectrodes, in 0.1 M $NaClO_4$, onset potentials were only -0.2 or -0.8 V vs. NHE, with Faradaic yields close to 100%. However, the turnover numbers were very low, possibly because of carbon deposits on the photoelectrodes.[32-34]

Various organic mediators were tested for CO_2 reduction on p-CdTe photocathodes. The best combination for methanol production was that using 15-crown-5 or 18-crown-6 in DMF-Bu$_4$NClO$_4$-5% H_2O, in which the current efficiencies for CO and methanol at ambient temperature were 85 and 13%, at a current density of about 9 mA/cm^2. The onset potential was extremely low, only -0.1 V vs. NHE, which was due to an anodic shift of 410 mV relative to that of the bare CdTe electrode. The mechanism of action of the crown ethers was proposed to be due to their adsorption on the electrode, displacing the adsorbed solvent molecules, and by coordinating the tetraalkylammonium cations. These then were adsorbed on the inner Helmholtz layer on the semiconductor. Electron transfer from the electrode to the tetraalkylammonium cation resulted in a tetraalkylammonium radical, which mediated electron transfer to CO_2.[35,36]

$$NR_4^+ + e^- = \cdot NR_4(ads) \tag{1}$$

$$\cdot NR_4(ads) + CO_2(ads) = NR_4^+ + CO_2^-(ads) \tag{2}$$

$$CO_2^-(ads) + H^+ + e^- + h\nu \rightarrow CO(ads) + OH^- \tag{3}$$

$$OH^- + H^+ \rightarrow H_2O \tag{4}$$

Until now, most photoelectrochemical CO_2 reduction systems suffered from low current densities, probably due to the high resistances of the semiconductor electrodes.

References

1. **Fujishima, A. and Honda, K.,** Electrochemical photolysis of water at a semiconductor electrode, *Nature,* 238, 37–38, 1972.

2. **Ulman, M.,** M.Sc. thesis, Weizmann Institute of Science, 1982.

3. **Habib, M. A. and Bockris, J. O' M.,** FT-IR spectrometry for the solid/solution interface, *J. Electroanal. Chem.,* 180, 287–306, 1984.

4. **Aurian-Blajeni, B., Habib, M. A., Taniguchi, I., and Bockris, J. O' M.,** The study of adsorbed species during the photoassisted reduction of carbon dioxide at a p-CdTe electrode, *J. Electroanal. Chem.,* 157, 399–404, 1983.

5. **Halmann, M.,** Photoelectrochemical reduction of aqueous carbon dioxide on p-type gallium phosphide in liquid junction solar cells, *Nature,* 275, 115–116, 1978.

6. **Halmann, M. and Aurian-Blajeni, B.,** Semiconductor electrolyte solar cells for the photoelectrochemical reduction of carbon dioxide to solar fuel, Proc. 1979 Photovoltaic Solar Energy Conf., Berlin (West), April 23–26, 1979, 682.

7. **Inoue, T., Fujishima, A., Konishi, S., and Honda, K.,** Photoelectrocatalytic reduction of carbon dioxide in aqueous suspensions of semiconductor powders, *Nature,* 277, 637–638, 1979.

8. **Ito, K., Ikeda, S., Yoshida, M., Ohta, S., and Iida, T.,** On the reduction products of carbon dioxide at a p-type gallium phosphide photocathode in aqueous electrolytes, *Bull. Chem. Soc. Jpn.,* 57, 583–584, 1984.

9. **Ikeda, S., Yoshida, M., and Ito, K.,** Photoelectrochemical reduction products of carbon dioxide at metal coated p-GaP photocathodes in aqueous electrolytes, *Bull. Chem. Soc. Jpn.,* 58, 1353–1357, 1985.

10. **Noda, H., Yamamoto, A., Ikeda, S., Maeda, M., and Ito, K.,** Influence of light intensity on photoelectroreduction of CO_2 at a p-GaP photocathode, *Chem. Lett.,* 1990, 1757–1760.

11. **Ikeda, S. and Ito, K.,** Artificial photosynthetic systems for carbon dioxide fixation, Proc. Int. Symp. Chemical Fixation of Carbon Dioxide, Nagoya, Japan, Dec. 2–4, 1991, 23–30.

12. **Taniguchi, I., Aurian-Blajeni, B., and Bockris, J. O'M.,** Photo-aided reduction of carbon dioxide to carbon monoxide, *J. Electroanal. Chem.,* 157, 179–1982, 1983.

13. **Taniguchi, I., Aurian-Blajeni, B., and Bockris, J. O' M.,** The mediation of the photoelectrochemical reduction of carbon dioxide by ammonium ions, *J. Electroanal. Chem.,* 161, 385, 1984.

14. **Taniguchi, I., Aurian-Blajeni, B., and Bockris, J. O' M.,** The reduction of carbon dioxide at illuminated p-type semiconductor electrodes in nonaqueous media, *Electrochim. Acta,* 29, 923–932, 1984.

15. **Yoneyama, H., Sigumura, K., and Kuwabata, S.,** Effects of electrolytes on the photoelectrochemical reduction of CO_2 at illuminated p-type CdTe and p-type InP electrodes in aqueous solutions, *J. Electroanal. Chem.,* 249, 143–153, 1988.

16. **Ikeda, S., Amakusa, S., Noda, H., Saito, Y., and Ito, K.,** Photoelectrochemical and electrochemical formation of methane from carbon dioxide at copper coated electrodes, *Proc. Electrochem. Soc., Photoelectrochemistry and Electrosynthesis on Semiconducting Materials,* Vol. 88–14, 1988, 130–136.

17. **Ikeda, S., Saito, Y., Yoshida, M., Noda, H., Maeda, M., and Ito, K.,** Photoelectrochemical reduction products of CO_2 metal coated p-GaP photocathodes in non-aqueous electrolytes, *J. Electroanal. Chem.,* 260, 335–345, 1989.

18. **Noda, H., Ikeda, S., Saito, Y., Nakamura, T., Maeda, M., and Ito, K.,** Photoelectrochemical reduction of carbon dioxide at metal-coated p-InP photocathodes, *Denki Kagaku,* 57, 1117–1120, 1989; *Chem. Abstr.,* 112, 65316h.

19. **Chu, C. Y., Peihai, W., Junfu, L., and Shiyong, Z.,** Catalysis of TiO_2 thin film for photoelectrochemical reduction of carbon dioxide, Proc. Int. Symp. Chemical Fixation of Carbon Dioxide, Nagoya, Japan, Dec. 2–4, 1991, 31–38.

20. **Liu, J.-F. and Baozhu, C. Y.,** Photoelectrochemical reduction of carbon dioxide on a p^+/p-Si photocathode in aqueous electrolyte, *J. Electroanal. Chem.,* 324, 191–200, 1992.

21. **Aurian-Blajeni, B., Halmann, M., and Manassen, J.,** Electrochemical measurements on the photoelectrochemical reduction of aqueous carbon dioxide on p-gallium phosphide and p-gallium arsenide semiconductor electrodes, *Solar Energy Mater.,* 8, 425–440, 1983.

22. **Ueda, J. and Uosaki, K.,** Photoelectrochemical fixation of CO_2 at semiconductor electrodes, Proc. Int. Symp. Chemical Fixation of Carbon Dioxide, Nagoya, Japan, Dec. 2–4, 1991, 343–346.

23. **Fujishima, A.,** Electrochemical carbon dioxide reduction using solar energy, Proc. Int. Symp. Chemical Fixation of Carbon Dioxide, Nagoya, Japan, Dec. 2–4, 1991, 11–18.

24. **Saeki, T., Hashimoto, K., and Fujishima, A.,** Electrochemical reduction of CO_2 on bi-metal electrodes — bimetal, compound semiconductor and highly dispersed metal electrodes, Proc. Int. Symp. Chemical Fixation of Carbon Dioxide, Nagoya, Japan, Dec. 2–4, 1991, 297–302.

25. **Bradley, M. G., Tysak, T., Graves, D. J., and Vlachopoulos, N. A.,** Electrocatalytic reduction of carbon dioxide at illuminated p-type silicon semiconducting electrode, *J. Chem. Soc. Chem. Commun.,* 1983, 349–350.

26. **Zafrir, M., Ulman, M., Zuckerman, Y., and Halmann, M.,** Photoelectrochemical reduction of carbon dioxide to formic acid, formaldehyde and methanol on p-gallium arsenide in an aqueous V(II)-V(III) chloride redox system, *J. Electroanal. Chem.,* 159, 373–389, 1983.

27. **Taniguchi, Y., Yoneyama, H., and Tamura, H.,** Photoelectrochemical reduction of carbon dioxide at p-type gallium phosphide electrodes in the presence of crown ether, *Bull. Chem. Soc. Jpn.,* 55, 2034–2039, 1982.

28. **Diaz, A. F. and Logan, J. A.,** Electroactive polyaniline films, *J. Electroanal. Chem.,* 111, 111–114, 1980.

29. **Noufi, R., Nozik, A. J., White, J., and Warren, L. F.,** Enhanced stability of photoelectrodes with electrogenerated polyaniline films, *J. Electrochem. Soc.,* 129, 2261–2265, 1982.

30. **Aurian-Blajeni, B., Taniguchi, I., and Bockris, J. O' M.,** Photoelectrochemical reduction of carbon dioxide using polyaniline-coated silicon, *J. Electroanal. Chem.,* 149, 291–293, 1983.

31. **Cabrera, C. R. and Abruña, H. D.,** Electrocatalysis of CO_2 reduction at surface modified metallic and semiconducting electrodes, *J. Electroanal. Chem.,* 209, 101–107, 1986.

32. **Beley, M., Collin, J.-P., Sauvage, J.-P., Petit, J.-P., and Chartier, P.,** Photoassisted electro-reduction of CO_2 on p-GaAs in the presence of Ni cyclam^{2+}, *J. Electroanal. Chem.,* 206, 333–339, 1986.

33. **Petit, J.-P., Chartier, P., Beley, M., and Sauvage, J.-P.,** Selective photoelectrochemical reduction of CO_2 to CO in an aqueous medium on p-GaP, mediated by Ni cyclam^{2+}, *New J. Chem.,* 11, 751–752, 1987.

34. **Petit, J.-P., Chartier, P., Beley, M., and Deville, J.-P.,** Molecular catalysts in photoelectrochemical cells. Study of an efficient system for the selective photoelectroreduction of CO_2: p-GaP or p-GaAs/Ni(cyclam)$^{2+}$, aqueous medium, *J. Electroanal. Chem.,* 269, 267–281, 1989.

35. **Bockris, J. O' M. and Wass, J. C.,** The photoelectrocatalytic reduction of carbon dioxide, *J. Electrochem. Soc.,* 136, 2521–2528, 1989.

36. **Bockris, J. O' M. and Gonzáles-Martin, A.,** The photoelectrochemical reduction of carbon dioxide, Proc. Int. Symp. Chemical Fixation of Carbon Dioxide, Nagoya, Japan, Dec. 2–4, 1991, 63–72.

NINE

Heterogeneous Photoassisted Reduction

Semiconductors may also be used as powdered substances, dispersed as slurries in aqueous or nonaqueous solvents, or as photocatalysts in gas-solid reactions. The particles of such materials have been considered as minute electrolysis cells.[1] Under illumination at wavelengths shorter than the band gap, these particles may cause photocatalytic reactions, including the reduction of carbon dioxide to carbon monoxide and to several organic compounds, such as formic acid and methanol. The simplicity of operating such photocatalytic reactions has stimulated considerable recent efforts to improve the efficiency and selectivity of such processes to desired products. In some cases, high quantum yields of formic acid were reported, and in a few cases also of methanol and ethanol.

AQUEOUS SUSPENSIONS

When the sizes of these particles were decreased to the dimensions of colloidal particles, quantum effects in light absorption and in chemical reactivity were discovered.[2] The quantization effects were explained to be due to the confinement of the charge carriers, electrons, and holes, in the very small particles. This confinement results in perturbations of the semiconductor band structure, causing increases in the effective band gaps, and leading to series of discrete states in the conduction and valence bands. The increased effective band gap results in enhanced redox potentials for the photoexcited electrons and holes. Thus, these electrons can perform reactions which are not possible with large-particle semiconductors. With CdSe colloids (diameter $D_p < 50$ Å, prepared from $Cd(ClO_4)_2$

and H_2Se gas in water-alcohol mixture at $-20°C$ and stabilized by SiO_2 colloids), CO_2 in aqueous solution was photoreduced to formic acid. No photoreduction of CO_2 was observed with CdSe of larger particle sizes.[3]

Photocatalytic reactions on semiconductor particles depend on a primary step of chemisorption. One tool to clarify the structure of carbon dioxide chemisorbed on TiO_2 is infrared absorption spectroscopy. In the case of reduced Pt/TiO_2, the absorbed species was proposed to form a bidentate carbonate structure, which was identified by its 1245 and 1673 cm^{-2} absorption bands,[4]

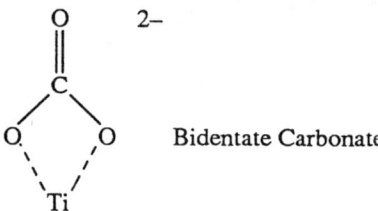 Bidentate Carbonate

Following the primary step of adsorption of carbon dioxide on the semiconductor particles, electrons and holes may transfer from the photoexcited semiconductor to surface-adsorbed molecules. Interaction of these excited surface molecules with the medium may result in the release of highly reactive intermediate species in solution. Some of these short-lived transient species, such as ·OH and ·O_2H radicals, can be stabilized by transforming them into *spin adducts* with much longer life times.[5] Spin adducts based on nitrones have been detected by electron spin resonance (ESR).[6,7] In experiments with illuminated suspensions of tungsten oxide in aqueous sodium hydrogen carbonate containing the spin trap PBN (α-phenyl-N-*tert*-butylnitrone), a spin adduct was detected, which was assigned as the ·CO_3^- radical anion.[8]

In most experiments on the photoassisted reduction of carbon dioxide, the main products have been formic acid or carbon monoxide. A much higher degree of reduction was observed by illuminating a single crystal of strontium titanate (111 crystal face), which was in contact with a platinum foil, in an atmosphere of carbon dioxide and water vapor. Methane was detected by gas chromatography, indicating an eight-electron reduction of carbon dioxide.[9]

With colloidal dispersions of platinized TiO_2 (prepared by acid hydrolysis of titanium tetraisopropoxide) in 1 M aqueous N_2CO_3 under illumination at $\lambda > 300$ nm, the only organic product was formaldehyde. Flash photolysis with a 337-nm N_2 laser of a similar dispersion of colloidal TiO_2 in 0.1 M Na_2CO_3 produced a short-lived transient absorption peak at 600 nm, decaying exponentially with a half-life of 0.9 μs, which was assigned as the carbonate radical anion,[10]

$$TiO_2 + h\nu \rightarrow e^- + h^+ \qquad (1)$$

Surface-adsorbed carbonate anions inject electrons into the holes,

$$CO_3^{2-} + h^+ \rightarrow ·CO_3^- \qquad (2)$$

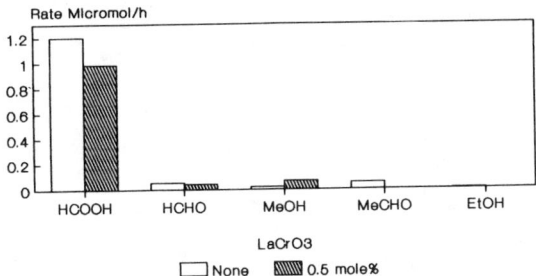

Fig. 1 Photoreduction in aqueous $SrTiO_3$. $LaCrO_3$ effect.

In the presence of even traces of oxygen, there may be charge transfer to form the neutral carbonate radical, which may decompose into surface-adsorbed CO and oxygen,

$$\cdot CO_3^- + O_2 \rightarrow \cdot CO_3 + O_2^- \qquad (3)$$

$$\cdot CO_3 \rightarrow CO_{ad} + O_2 \qquad (4)$$

The adsorbed CO presumably was further reduced on the semiconductor, releasing formaldehyde. Methanol formation was not detected in this reaction.[10] Formaldehyde production was also reported in a study of the effect of different metals coated on TiO_2. Highest yields of HCHO were obtained by illumination of CO_2-saturated aqueous suspensions of Hg-coated TiO_2.[11]

In efforts to enhance the yields of CO_2 photoreduction products, the effect of surface treatment with transition metal oxide additives on powdered strontium titanate was tested. The organic products observed included formic acid, formaldehyde, methanol, acetaldehyde, and ethanol. Results for the production rates of the organic products with either plain $SrTiO_3$ or $LaCrO_3$ (0.5%) on $SrTiO_3$ are presented in Figure 1,[12] showing that the lanthanum chromite surface doping caused a decrease in the production of formic acid, but enhanced the formation of methanol. With the same $LaCrO_3/SrTiO_3$ catalyst, the photoreduction of CO_2 was compared in water and in 0.1 M $LiHCO_3$. As shown in Figure 2, the production rates of formaldehyde and methanol were enhanced in the $LiHCO_3$ medium, while HCOOH production was faster in water as the medium.[12]

Metal loading of TiO_2 considerably enhanced its efficiency as photocatalyst in aqueous suspensions in the reduction of carbon dioxide to methane. Palladium loading increased the activity by more than ten times relative to bare TiO_2. The order of activity of several metal deposits in accelerating methane production was Pd > Rh > Pt > Ru.[13]

In a comparison of the effectiveness of various semiconductors, highest yield of photoreduction of CO_2 was obtained with SiC. After 7 hours of illumination, the concentrations of formaldehyde and of methanol were with TiO_2 1.1 and 0.23

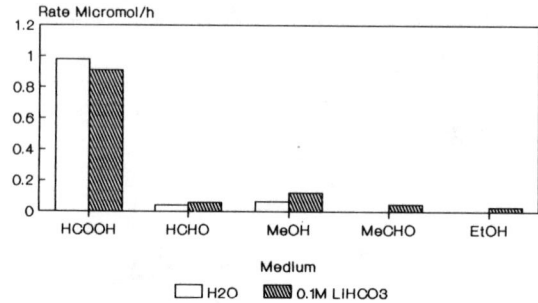

Fig. 2   Photoreduction in aqueous SrTiO$_3$/LaCrO$_3$. LiHCO$_3$ effect.

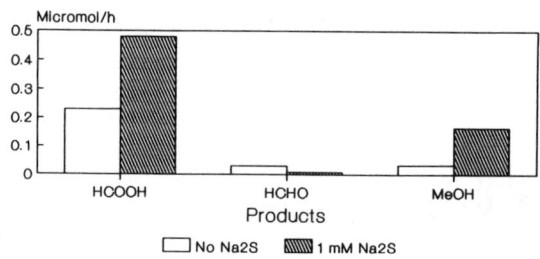

Fig. 3   Photoreduction in aqueous TiO$_2$/RuO$_2$/Cr. Sulfide effect.

mM, with ZnO 1.2 and 0.35 mM, with CdS 2.0 and 1.2 mM, with GaP 1.0 and 1.2 mM, with SiC 1.0 and 5.4 mM, and with WO$_3$ 0.0 and 0.0 mM. The yields correlated with the level of the conduction band of the semiconductor used, the most negative conduction band level being that of SiC, about -1.6 V vs. NHE. Using SiC as photocatalyst, the quantum yields for production of formaldehyde and methanol were 0.05 and 0.45%, respectively.[14] In the photoassisted reduction of carbonated aqueous suspensions of several powdered semiconductors by illumination with high-pressure Hg lamps, the productions of formaldehyde and methanol was measured. The order of activity of the semiconductors used was SrTiO$_3$ > WO$_3$ > TiO$_2$, resulting in absorbed energy conversion efficiencies of 6, 5.9, and 1.2%, respectively.[15]

In the photoassisted reduction of carbon dioxide passed through aqueous suspensions of TiO$_2$ in 0.1 M Li$_2$CO$_3$, during 47 h of illumination with a 75-W high-pressure Hg lamp, the production rates of formic acid and methanol were considerably increased by the presence of 1 mM Na$_2$S, as shown in Figure 3. Presumably the sulfide ions serve as sacrificial hole traps or as electron donors.[16]

The high band gap semiconductors barium titanate and lithium niobate, doped with the rare earths europium, neodynium, and samarium, in aqueous carbonated

Fig. 4 Photochemical solar collector.

suspensions, were tested as photocatalysts for the reduction of carbon dioxide. The yields of formic acid and formaldehyde were markedly enhanced by the presence of the rare earth dopants. The order of effectiveness of the rare earths was $Nd_2O_3 > Sm_2O_3 \gg Eu_2O_3$.[17]

For testing the photoassisted reduction of carbon dioxide under illumination with natural sunlight, a photochemical solar collector was built. The device consisted of a rectangular aluminum plate, which was painted with an aqueous slurry of the powdered semiconductor to be studied, immersed in water through which CO_2 was bubbled, and which was surrounded by an aluminum frame, and exposed to sunlight through a coverplate of glass or methyl methacrylate (1×0.5 m^2 area). The outflowing gas was passed through a series of two to three traps cooled to 0°C, to collect the organic reduction products (see Figure 4).[18]

The products were mainly formic acid, as well as smaller amounts of formaldehyde, methanol, ethanol, and acetaldehyde. Results for experiments made in carbonated aqueous 0.1 M Li_2CO_3 are presented in Figure 5, in which the production rates of the organic products are expressed on a log scale as $10^3 \times$ μmol/ KJ (normalized by the light dose). The semiconductor materials tested were TiO_2, plain $SrTiO_3$, and $SrTiO_3$ doped with 0.5 mol% $LaCrO_3$. The TiO_2 used was the

**Photoreduction of Carbon Dioxide
in Photochemical Solar Collector**

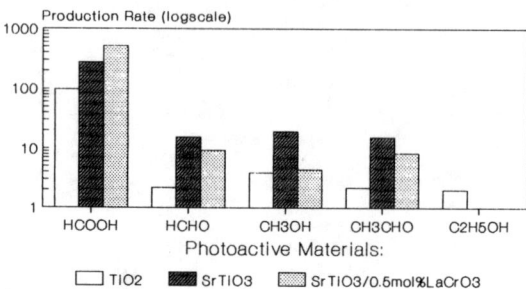

Fig. 5 Solar photoreduction. Catalyst effect.

least effective of the three materials. The predominant product with all photoactive agents was formic acid. Methanol production was highest with the plain $SrTiO_3$.[18]

With a CdS-ZnS mixture (1.4:1.0 ratio) suspended in 0.5 M K_2CO_3-0.1 M Na_2S, thermostatted at 61°C, and under argon as carrier gas, illumination with a 150-W xenon lamp caused the production of methanol and formaldehyde at rates of 1.5 $\times$ 10^{-9} and 0.2 $\times$ 10^{-9} mol h^{-1} cm^{-2} (per illuminated area). Formic acid was not detected.[19]

While formic acid was the predominant product of the photoreduction of CO_2 in aqueous solutions in the presence of TiO_2/Pt, other organic products detected were formaldehyde, methanol, acetaldehyde, and ethanol, the accumulation of which in long-term illumination is presented in Figure 6.[20]

With platinized TiO_2 doped with Cr or Mn (0.5 atom %) suspended in 1 M Na_2CO_3 in a nitrogen atmosphere, illuminated with a medium-pressure mercury lamp, a photoreduction took place yielding carbon as well as smaller amounts of formaldehyde and methanol.[21,22] In similar experiments, but with TiO_2 (rutile) coated with Fe^{2+} or Co^{2+} phthalocyanines (2% surface coverage suspended in aqueous sodium carbonate under nitrogen) enhanced yields of methanol and formaldehyde were reported.[23]

The presence of hydrogen sulfide was found to enhance the photoassisted reduction of carbon dioxide to formic acid and formaldehyde in aqueous suspensions of amorphous $n-Bi_2S_3$ or n-CdS semiconductors.[24]

Very high quantum yields, up to 80%, were reported for CO_2 reduction to HCOOH in aqueous media, using microcrystalline ZnS, which was stabilized by colloidal SiO_2, with sulfite as a hole acceptor. The mechanism proposed involved a two-electron reduction.[25] Colloidal ZnS was found also to catalyze the photoreduction of carbon dioxide by methanol or 2-propanol. 2-Propanol was much more efficient than methanol in promoting the production of formic acid, resulting in a quantum yield of up to 80%.[26] In more recent studies with ZnS, much lower quantum yields were observed, maximally reaching 30%. The quantum yield was found to increase with the $[Zn^{2+}]/[S^{2-}]$ ratio used in the preparation of the ZnS colloids.[27] With suspensions of powdered ZnS in water/2,5-dihydrofuran, the

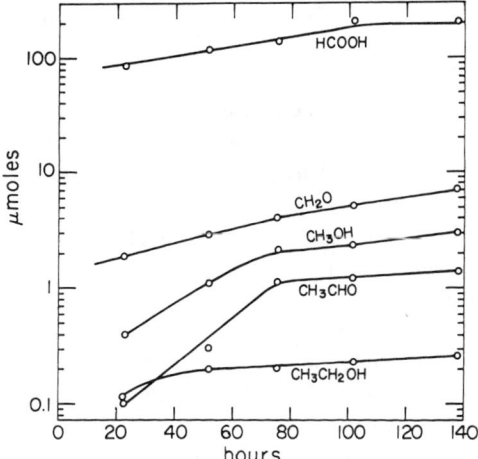

Fig. 6 Photoreduction in aqueous TiO₂/Pt. Products evolution.

photoreduction of carbon dioxide to formic acid occurred with a quantum yield of only 0.1% (at $\lambda = 300$ nm). The 2,5-dihydrofuran functioned as reducing agent, and was itself oxidized to dehydro dimers. The mechanism proposed involved two-electron transfer from zinc sulfide to adsorbed hydrogen carbonate.[28] Low quantum yields (0.3%) for CO_2 reduction were also found using Pd/TiO₂ dispersions.[29]

The above widely different results with ZnS may possibly be understood by different methods of preparing the photocatalyst. Defect-free colloidal ZnS crystallites were precipitated by mixing equivalent amounts of aqueous ZnSO₄ and Na₂S at 0°C under argon. With these quantum crystallites, in an aqueous mixture of Na₂S and NaH₂PO₂ saturated with CO_2 at pH 7, UV illumination at >290 nm resulted in production of formic acid, carbon monoxide, and hydrogen at initial rates of 75, 1.7, and 86 μmol h⁻¹, respectively. The apparent quantum yield for formic acid was $\Phi_{1/2HCOOH} = 0.24$ at 313 nm. In this reaciton, hypophosphite ions served as sacrificial electron donors, while sulfide ions served as suppressors of surface defects. Since these quantized microcrystallites of ZnS had in their reflectance spectra a steep onset, at shorter wavelength than with ordinary preparations of ZnS, these microcrystallites presumably have a lower density of surface defects, which could be the reason for their improved effectiveness for CO_2 reduction.[30,31]

With colloidal CdS microcrystallites, the selective photoreduction of CO_2 to CO was achieved even by visible light, in dimethylformamide solution containing 1 v/v% water, with triethylamine as sacrificial electron donor. These microcrystallites were prepared from Cd(ClO₄)₂ and H₂S in DMF solution, and were shown by high-resolution electron microscopy to range in size from 3 to 5 nm in diameter. The apparent quantum yield was $\Phi_{1/2CO} = 0.098$ at $\lambda > 400$ nm. Ordinary powdered CdS was inactive.[32] Size quantization effects were proposed to account for the photocatalytic activity, in analogy to the above results with microcrystallites of ZnS.[30,31]

While CO_2 reduction in water as solvent yielded mainly formic acid, the addition of tetraalkylammonium salts modified the semiconductor surface. In water alone as medium, with colloidal CdS as photocatalyst, CO_2 was photoreduced to formic acid, methanol, and formaldehyde in the ratio 87:75:11, with a total quantum yield of 0.035%. In the presence of tetramethyl ammonium chloride, with the addition of sulfite to the medium, the product composition changed to that of formate and glyoxylate, in the ratio 22:12, with a total quantum yield of 0.1%.[33]

Much improved yields of formic acid were obtained by using Pd-colloid stabilized by β-cyclodextrin. In aqueous $NaHCO_3$ as medium, with deazariboflavin as photosensitizer, methylviologen (MV^{2+}) as electron relay, and oxalate as sacrificial electron donor, the quantum yield with visible light reached 110%.[34]

Visible light photoreduction of CO_2 to CH_4, C_2H_4, and C_2H_6 in aqueous solutions was obtained using Ru or Os colloids as catalysts. In one system, $Ru(II)(bpy)_3^{2+}$ was the photosensitizer, triethanolamine (TEOA) the electron donor, and one of several bipyridinium compounds the mediators (charge relays). With N,N'-bis-(3-sulfonatopropyl)-3,3'-dimethyl-4,4'-bipyridinium as mediator, in 0.1 M aqueous bicarbonate solution (pH 7.8), with Ru colloid as catalyst, the quantum yields Φ were for H_2: 2.6×10^{-3}, for CH_4: 5.7×10^{-4}, and for C_2H_4: 1.9×10^{-5}. Even lower yields were obtained with the Os colloid. In a second system, Ru(II)-tris(bipyrazine), $Ru(bpz)_3^{2+}$, was the photosensitizer, TEOA the electron donor, and colloidal Ru the catalyst, in 0.05 M $NaHCO_3$-ethanol (2:1, pH 7.8) as medium. With this sensitizer, the production of H_2 was inhibited. Values of Φ were CH_4: 4.0×10^{-4}, C_2H_4: 7.5×10^{-5}, and C_2H_6: 4×10^{-5}. The turnover number TN based on the $Ru(bpz)_3^{2+}$ sensitizer was 15. This sensitizer absorbs strongly in the visible region, with $\lambda_{max} = 443$ nm, $\epsilon = 15000$ M^{-1} cm^{-1}. It has a long-lived excited state ($\tau = 1.04$ μs). Its reduced photoproduct, $Ru(bpz)_3^{2+}$, is a strong reducing agent, with $E°$ $[Ru(bpz)_3^+/Ru(bpz)_3^{2+}] = -0.86$ V vs. SCE, thus adequate to reduce CO_2 to CH_4.[35,36]

Selective photoreduction of carbon dioxide to formate was obtained using suspensions or colloids of Pd/TiO_2, stabilized by polyvinyl alcohol and β-cyclodextrin, in the presence of sacrificial oxalate. Quantum yields for formate production were about 1.4%.[37]

A marked dependence on the size of particles was found in a study of powdered SiC. Thus, with 100-mesh SiC, the photoreduction of carbon dioxide in aqueous media yielded only traces of methanol, and no detectable amounts of ethanol. However, using 1000-mesh SiC, in aqueous sulfuric acid (pH 2.9), during 8 h of illumination, ethanol and methanol were produced in amounts of 18 and 12 μmol, respectively.[38] In a further study, these authors tested the effect of metal-loading on the 1000-mesh SiC particles.[39] Highest concentrations of organic products were obtained with SiC/Pd, with which CO_2 was photoreduced to HCOOH, HCHO, methanol, acetaldehyde, and ethanol, in an overall energy conversion efficiency of 0.013%. This efficiency was defined as

$$\text{Effic. } (\%) = \frac{100(\text{Heat of Combustion of Products})}{(\text{Incident Light Energy})}$$

With a mixture of p-SiC (325 mesh) and copper powder (100 mesh) in 0.5 M aqueous $KHCO_3$ (pH 5), CO_2 photoreduction resulted in production rates of methane, ethylene, and ethane of 1.49, 0.39, and 0.23 μlit h^{-1}.[40]

While the illumination of carbon dioxide in the presence of aqueous suspensions of TiO_2 alone is quite ineffective for the reduction of CO_2, appreciable reduction was attained when using mixtures of TiO_2 and metallic copper. With TiO_2 (0.5 g) and Cu powder (0.3 g; mean particle diameter 200 μm) dispersed in CO_2-saturated water, illumination with a xenon lamp resulted in the production of small amounts of CO, formic acid, formaldehyde, and methanol, in concentrations gradually increasing in time. The concentrations of formaldehyde and methanol reached a maximum at about 9 to 10 min, and then declined. Even larger amounts of methanol were produced when the TiO_2–Cu mixture was illuminated in aqueous sodium bicarbonate. Free metallic copper was required for this photocatalyzed reduction of CO_2 in TiO_2 powder suspensions. No reduction was observed on copper supported on TiO_2. Presumably, the metallic copper acts both as an effective catalyst for the reduction (as in the electrochemical reduction on copper electrodes) and as a sacrificial reducing species, being oxidized by the positive holes generated on the illuminated semiconductor.[41]

An interesting approach to enhance the efficiency of reduction of carbon dioxide has been to use a semiconductor which is also a sacrificial hole trap. With an suspension of hydrous cuprous oxide ($Cu_2O.xH_2O$), selective and enhanced production of methanol and formaldehyde was achieved, reaching maximal concentrations of 24.0 and 3.5 μmol l^{-1} after about 45 min.[42]

GAS-SOLID REACTIONS

A variety of metal oxide catalysts were tested for the gas phase photoassisted reduction of CO_2 by H_2 in the absence and presence of water vapor under visible light. In experiments at 30°C and 1-atm pressure, highest formate production was observed with $Pt/LaNiO_3$ as catalyst, followed by Co–Mo/Al_2O_3 and by α-Fe_2O_3. In the absence of water vapor in the dark, $CO_2 + H_2$ (1:2) over all catalysts produced CO as the major product, and small amounts of CH_4. In the presence of water vapor and under illumination, yields of HCOOH became appreciable, using Zn-Fe-oxide or Co–Mo/Al$_2$O$_3$ as catalysts.[43,44]

Using a highly dispersed CeO_2–TiO_2 catalyst (0.5 wt% CeO_2, prepared by co-precipitation from titanium(IV) sulfate and cerium(III) nitrate), under illumination with visible light ($\lambda > 370$ nm), the gas-solid phase photolysis of a H_2O–CO_2 mixture was performed. With initial gas pressures of 100 torr CO_2 and 25 torr H_2O, at 25°C, the products were H_2, O_2, and CH_4, formed at initial rates (during the first hour) of 36, 14, and 3.4 μmol (g catalyst)$^{-1}$ h^{-1}. When the catalyst had been pretreated in H_2, the yield of methane was linearly related to the CO_2 pressure.

It was thus proposed that the primary reaction was the photoassisted dissociation of water to hydrogen, and that some of the hydrogen carried out a methanation reaction on the carbon dioxide. Considering each methane molecule to contain four hydrogen atoms, the maximal quantum yield for the formation of hydrogen (H_2 + $2CH_4$) was estimated to be 0.05%.[46]

The photoassisted reduction of CO_2 by NH_3 was carried out in a gas-solid reaction at 5°C, using pre-adsorbed ammonia on silica-supported iron (prepared by impregnation of silica gel with ferric nitrate). The catalyst, which was pretreated with H_2, contained 5 wt% Fe/SiO_2. The iron was shown by X-ray diffraction to be in the form of Fe_3O_4. The major products were CO, CH_4, H_2, and N_2. The quantum yield of CO formation increased with the amount of adsorbed NH_3, reaching a plateau of $\Phi = 0.5$ at higher values of adsorbed NH_3. It was proposed that co-adsorbed CO_2 and NH_3 were involved in the initiation of the photolysis. This was followed by the reaction of adsorbed CO with adsorbed H atoms, leading to the production of methane.[46]

References

1. **Bard, A. J.,** Photoelctrochemistry and heterogeneous photocatalysis at semiconductors, *J. Photochem.*, 10, 59–75, 1979.
2. **Rossetti, R., Nakahara, S., and Brus, L. E.,** Quantum size effects in the redox potentials, resonance Raman spectra, and electronic spectra of CdS crystallites, *J. Chem. Phys.*, 79, 1086–1088, 1983.
3. **Nedeljkovic, J. M., Nenandovic, M. T., Micic, O. I., and Nozik, A. J.,** Enhanced photoredox chemistry in quantized semiconductor colloids, *J. Phys. Chem.*, 90, 12–13, 1986.
4. **Tanaka, K. and White, J. M.,** Dissociative adsorption of CO_2 on oxidized and reduced Pt/TiO_2, *J. Phys. Chem.*, 86, 3977–3980, 1982.
5. **Evans, C. A.,** Spin trapping, *Aldrichim. Acta*, 12, 23–29, 1979.
6. **Harbour, J. R. and Hair, M. L.,** *J. Phys. Chem.*, Superoxide generation in the photolysis of aqueous cadmium sulfide dispersions. Detection by spin trapping, 81, 1791–1793, 1977.
7. **Jaeger, C. D. and Bard, A. J.,** *J. Phys. Chem.*, Spin trapping and electron spin resonance detection of radical intermediates in the photodecomposition of water at TiO_2 particulate systems, 83, 3146–3152, 1979.
8. **Aurian-Blajeni, B., Halmann, M., and Manassen, J.,** Radical generation during the illumination of aqueous suspensions of tungsten oxide in the presence of methanol, sodium formate and bicarbonate. Detection of spin trapping, *Photochem. Photobiol.*, 35, 157–162, 1982.
9. **Hemminger, J. C., Carr, R., and Somorjai, G. A.,** The photoassisted reaction of gaseous water and carbon dioxide adsorbed on the $SrTiO_3$ (111) crystal face to form methane, *Chem. Phys. Lett.*, 57, 100–104, 1978.
10. **Chandrasekaran, K. and Thomas, J. K.,** Photochemical reduction of carbonate to formaldehyde on TiO_2 powder, *Chem. Phys. Lett.*, 99, 7–10, 1983.
11. **Tennakone, K.,** Photoreduction of carbonic acid by mercury coated n-titanium oxide, *Solar Energy Mater.*, 10, 235–238, 1984.

12. **Ulman, M., Tinnemans, A. H. A., Mackor, A., Aurian-Blajeni, B., and Halmann, M.**, Photoreduction of carbon dioxide to formic acid, formaldehyde, methanol, acetaldehyde and ethanol using suspensions of strontium titanate with transition metal additives, *Int. J. Solar Energy*, 1, 213–222, 1982.

13. **Ishitani, O. and Ibusuki, T.**, Metal-loaded TiO_2 photocatalyzed reduction of CO_2 to hydrocarbons, Abstr. 16th Int. Conf. on Photochemistry, Paris, July 28–August 2, 1991, VII–1.

14. **Inoue, T., Fujishima, A., Konishi, S., and Honda, K.**, Photoelectrocatalytic reduction of carbon dioxide in aqueous suspensions of semiconductor powders, *Nature*, 277, 637–638, 1979.

15. **Aurian-Blajeni, B., Halmann, M., and Manassen, J.**, Photoreduction of carbon dioxide and water into formaldehyde and methanol on semiconductor materials, *Solar Energy*, 25, 165–170, 1980.

16. **Halmann, M., Katzir, V., Borgarello, E., and Kiwi, J.**, Photoassisted carbon dioxide reduction on aqueous suspensions of titanium dioxide, *Solar Energy Mater.*, 10, 85–93, 1984.

17. **Ulman, M., Aurian-Blajeni, B., and Halmann, M.**, Photoassisted carbon dioxide reduction to organic compounds using rare earth doped barium titanate and lithium niobate as photoactive agents, *Israel J. Chem.*, 22, 177–179, 1982.

18. **Halmann, M., Ulman, M., and Aurian-Blajeni, B.**, Photochemical solar collector for the photoassisted reduction of aqueous carbon dioxide, *Solar Energy*, 31, 429–431, 1983.

19. **Halmann, M. and Zuckerman, K.**, Photoassisted reduction of carbon and nitrogen compounds with semiconductors, in *Homogeneous and Heterogeneous Photocatalysis*, Pelizzetti, E., Serpone, N., Eds., (NATO ASI Ser., Sr. C., Vol. 174), D. Reidel, Dordrecht, The Netherlands, 1986, 521–532.

20. **Ulman, M.**, M.Sc. thesis, Weizmann Instituted of Science, 1982.

21. **Rophael, M. W. and Malati, M. A.**, The photocatalysed reduction of aqueous sodium carbonate to carbon using platinised titania, *J. Chem. Soc. Chem. Commun.*, 1987, 1418–1420.

22. **Rophael, M. W. and Malati, M. A.**, The photocatalyzed reduction of aqueous sodium carbonate using platinized titania, *Photochem. Photobiol.*, 46, 367–377, 1989.

23. **Khalil, L. B., Youssef, N. S., Rophael, M. W., and Moawad, M. M.**, Reduction of aqueous carbonate photocatalysed by treated semiconductors as an application of solar energy conversion, Abstr., 8th Int. Conf. Photochemical Conversion and Storage of Solar Energy, Palermo, Italy, July 15–20, 1990, 214.

24. **Aliwi, S. M. and Al-Jubori, K. F.**, Photoreduction of carbon dioxide by metal sulfide semiconductors in presence of hydrogen sulfide, *Solar Energy Mater.*, 18, 223–229, 1989.

25. **Henglein, A.**, Catalysis of photochemical reactions by colloidal semiconductors, *Pure Appl. Chem.*, 56, 1215–1224, 1984.

26. **Henglein, A., Gutiérrez, M., and Fischer, Ch.-H.**, Photochemistry of colloidal metal sulfides. VI. Kinetics of interfacial reactions at zinc sulfide particles, *Ber. Bunsenges. Phys. Chem.*, 88, 170–175, 1984.

27. **Inoue, H., Torimoto, T., Sakata, T., Mori, H., and Yoneyama, H.**, Effects of size quantization of zinc sulfide microcrystallites on photocatalytic reduction of carbon dioxide, *Chem. Lett.*, 1990, 1483–1486.

28. **Kisch, H. and Twardzik, G.**, Zinc sulfide catalyzed photoreduction of carbon dioxide, *Chem. Ber.*, 124, 1161–1162, 1991.

29. **Albers, P. and Kiwi, J.**, Photochemical generation of formate via HCO_3^-/CO_2 reduction on Pd dispersions, *New J. Chem.*, 14, 135–139, 1990.

30. **Yanagida, S. and Kanemoto, M.,** Effective photoreduction of carbon dioxide to formate catalyzed by defect-free ZnS quantum crystallites in water, Proc. Int. Symp. Chemical Fixation of Carbon Dioxide, Nagoya, Japan, Dec. 2–4, 1991, 359–364.

31. **Kanemoto, M., Shiragami, T., Pac, C. J., and Yanagida, S.,** Semiconductor photocatalysis. Effective photoreduction of carbon dioxide catalyzed by ZnS quantum crystallites with low density of surface effects, *J. Phys. Chem.,* 96, 3521–3526, 1992.

32. **Kanemoto, M., Ishihara, K.-I., Wada, Y., Sakata, T., Mori, H., and Yanagida, S.,** Visible light induced effective photoreduction of CO_2 to CO catalyzed by colloidal CdS microcrystallites, *Chem. Lett.,* 1992, 835–836.

33. **Eggins, B. R., Irvine, J. T. S., Murphy, E. P., and Grimshaw, J.,** Formation of two-carbon acids from carbon dioxide by photoreduction on CdS, *J. Chem. Soc. Chem. Commun.,* 1988, 1123–1124.

34. **Mandler, D. and Willner, I.,** Effective photo-reduction of CO_2/HCO_3^- to formate using visible light, *J. Am. Chem. Soc.,* 109, 7884, 1987.

35. **Maidan, R. and Willner, I.,** Photoreduction of CO_2 to CH_4 in aqueous solutions using visible light, *J. Am. Chem. Soc.,* 108, 8100–8101, 1986.

36. **Willner, I., Maidan, R., Mandler, D., Dürr, H., Dörr, G., and Zengerle, K.,** Photosensitized reduction of CO_2 to CH_4 and H_2 in the presence of ruthenium and osmium colloids: strategies to design selectivity of products distribution, *J. Am. Chem. Soc.,* 109, 6080–6086, 1987.

37. **Goren, Z., Willner, I., Nelson, A. J., and Frank, A. J.,** Selective photoreduction of CO_2/HCO_3^- to formate by suspensions and colloids of Pd-TiO_2, *J. Phys. Chem.,* 94, 3784–3790, 1990.

38. **Yamamura, S., Kojima, H., Iyoda, J., and Kawai, W.,** Formation of ethyl alcohol in the photocatalytic reduction of carbon dioxide by SiC and ZnSe/metal powders, *J. Electroanal. Chem.,* 225, 287–290, 1987.

39. **Yamamura, S., Kojima, H., Iyoda, J., and Kawai, W.,** Photocatalytic reduction of carbon dioxide with metal-loaded SiC powders, *J. Electroanal. Chem.,* 247, 333–337, 1988.

40. **Cook, R. L., MacDuff, R. C., and Sammells, A. F.,** Photoelectrochemical carbon dioxide reduction to hydrocarbons at ambient temperature and pressure, *J. Electrochemical Soc.,* 135, 3069–3070, 1988.

41. **Hirano, K., Inoue, K., and Yatsu, T.,** Photocatalyzed reduction of CO_2 in aqueous TiO_2 suspension mixed with copper powder, *Photochem. Photobiol. A,* 64, 255–258, 1992.

42. **Tennakone, K., Jayatissa, A. H., and Punchihewa, S.,** Selective photoreduction of carbon dioxide to methanol with hydrous cuprous oxide, *J. Photoch. Photobiol. A,* 49, 369–375, 1989.

43. **Vijayakumar, K. M. and Lichtin, N. N.,** Reduction of CO_2 by H_2 and water vapor over metal oxides assisted by visible light, *J. Catal.,* 90, 173–177, 1984.

44. **Lichtin, N. N., Vijayakumar, K. M., and Rubio, B. I.,** Photoassisted reduction of CO_2 by H_2 over metal oxides in the absence and presence of water vapor, *J. Catal.,* 104, 246–251, 1987.

45. **Ogura, K., Kawano, M., Yano, J., and Sakata, Y.,** Visible-light-assisted decomposition of H_2O and photomethanation of CO_2 over CeO_2-TiO_2 catalyst, *J. Photochem. Photobiol. A.,* 66, 91–97, 1992.

46. **Ogura, K., Seno, A., and Kawano, M.,** Photo-assisted catalytic reduction of CO_2 with pre-adsorbed ammonia on silica supported iron, *J. Mol. Catal.,* 73, 225–235, 1992.

TEN

Bioconversion

Natural photosynthesis in green plants achieves carbon dioxide fixation on a global scale. However, the efficiency of solar energy conversion in plant production under optimal growth conditions is only 5 to 6%. Under field conditions, even high-yielding crops (such as maize, bulrush millet, or sugarcane) convert solar energy into plant material with a maximal efficiency of 1 to 2%. Most major crops and forests achieve much lower efficiencies. The global average efficiency has been estimated as 0.15%.[1] There exist thus considerable interest in other non-photosynthetic pathways for carbon dioxide fixation. One of these is that performed by the bacterial methanogens.[2,3]

BACTERIAL METHANOGENS

Non-photosynthetic carbon dioxide fixation occurs widely in nature — by methanogenic bacteria. These are obligate anaerobes, growing in freshwater and marine sediments, in peats, the gut of animals, sewage sludge, and manure piles.

In a study of the assimilation of CO_2 during the methylotrophic growth of bacteria, three different pathways of C_1 metabolism were identified. About 80% of cell carbon was formed from CO_2 by the ribulose biphosphate pathway, about 50% by the serine pathway, and about 15% by a hexulose phosphate pathway.[4] Further evidence indicated that the fixation of carbon dioxide may not involve the Calvin cycle, and that there occurs a direct synthesis of acetic acid, via coenzyme-bound carbon units and acetyl-CoA. These methanogenic bacteria grow optimally at temperatures between 20 and 95°C. They can use $CO_2 + H_2$ as their only sources of carbon and energy.[5]

CO_2 fixation by methanogenic bacteria has been proposed as a practical approach for the upgrading of waste gases from blast furnaces, to be used as fuel for steam boilers. By methanation of the carbon oxides, their caloric value may be significantly increased. Thus, with hydrogen as reducing agent, and mixtures of carbon monoxide and carbon dioxide as carbon source, methane was produced by thermophilic methanogens in a column bioreactor having 200-ml volume of media and supporters, operated at 55°C and pH 7.4. At a gas recirculation rate of 18 l/h, the rates of H_2 consumption and CH_4 production were 1380 and 300 mmol l^{-1} d^{-1}. The caloric value of the gas mixture, which originally had the composition of blast furnace gas, increased by this methanation from 755 to 6420 kcal mol^{-1}.[6]

Using a mixture of the components of low-BTU synthesis gas, CO, CO_2, and H_2, a completely biocatalytic conversion to the high-BTU methane was achieved using a mixture of cultures of three bacteria. The photosynthetic bacterium *Rhodospirillum rubrum* was applied to carry out the water gas shift reaction,

$$CO + H_2O \rightarrow H_2 + CO_2 \tag{1}$$

resulting in 100% conversion. Simultaneously, a mixture of two methanogens, *Methanobacterium formicicum* (which provides a high rate of hydrogen uptake, but is inhibited by CO), and *Methanosarcina barkeri* (which has a smaller rate of hydrogen uptake, but is more tolerant of CO), converted CO_2 to methane,

$$4H_2 + CO_2 \rightarrow CH_4 + 2H_2O \tag{2}$$

Using a trickle-bed reactor (a packed-bed tower), the methane yield was 83% of the theoretical required by the above methanation equation. The productivity of methane achieved at flow rates above 300 ml/h was 3.4 mmol l^{-1} h^{-1}, and the mass-transfer coefficient was 780 h^{-1}.[7]

While the conventional gas phase catalytic methods for methanation of CO_2 require temperatures of 300 to 700°C and pressures of 3 to 20 atm, and are sensitive to catalyst poisoning, e.g., by sulfur compounds (see Chapter 5), the biological conversion with the above triculture system could be operated at 37°C, was not affected by the presence of sulfur compounds, and was less sensitive to variations in the composition of the feedstock gases.

BIOCONVERSION TO METHANOL

An attractive option for the production of useful fuel could be to follow up on the methanation of synthesis gas with the bioconversion of methane to methanol. This has been achieved both with cultures of certain bacteria, and with cell-free extracts of the enzyme methane monooxygenase. Covalently immobilized cells of *Methylosinus trichosporium* were used both in batch and continuous culture to oxidize methane in high yield to methanol.[8,9] DEAE-cellulose linked cells of *M. trichosporium* were found to have highly specific methane monooxidase activity, enabling a methane oxidation rate of 66 μmol/h-mg cells. However, addition of sodium formate (20 mM) in the feedstream was required in order to maintain sustained methanol production.[9] Methanol was also produced by incubating a 1:1 mixture of methane and oxygen with the soluble enzyme methane oxygenase obtained by centrifuging the cell homogenate of *Methylobacterium organophylum*. The rate of methanol formation was 93 nmol min^{-1} mg protein^{-1}.[10] With whole cell cultures of *Methylosinus trichosporium*, there was production of methanol by oxidation of methane, with a yield of 30% based on the methane utilized. The methanol yield was enhanced by the presence of chloride ions and of hydrogen in the medium. The system using cell suspensions of *M. trichosporium* was suggested to be advantageous over enzyme preparations, as it avoids the need for expensive cofactors and enzyme purification.[8]

ENZYME-CATALYZED REACTIONS

Many enzymes are active in nature in catabolic processes, causing decarboxylation from organic compounds. Since these enzymes often act reversibly, they may be used in artificial systems for carboxylation, i.e., insertion of CO_2 into C-H bonds. This has been applied to the photosynthesis of formic, malic, aspartic, and other carboxylic acids, using photosensitized regeneration of NADPH, with a bipyridinium electron relay system, in the presence of the appropriate enzymes. Quantum yields for production of formate from CO_2 and for malic acid from pyruvic acid reached 1.6 and 1.9%, respectively.[11,12]

In a reversal of the natural action of the enzyme formate dehydrogenase (which catalyzes the conversion of formate to CO_2), using the methyl viologen redox system,

$$MV^{2+} + e^- = \cdot MV^+ \tag{3}$$

as mediator, the reduction of CO_2 to HCOOH was driven by either an electrochemical reaction, on graphite foil, or photoelectrochemically, on illuminated p-InP, using an aqueous phosphate buffer. At a potential of only $+0.05$ V vs. NHE, and a current density of 0.6 mA/cm^2, the current efficiencies reached 80 to 93%. However, the enzyme was unstable under the conditions of the reaction.[13]

The enzyme isocitrate dehydrogenase (ICDH) was used as electrocatalyst, with methylviologen (MV^{2+}) as mediator, to reverse the *in vivo* metabolic oxidation of isocitric acid to CO_2 and oxoglutaric acid,

$$\text{Oxoglutaric acid} + CO_2 \leftrightarrow \text{Isocitric acid} \tag{4}$$

With a glassy carbon electrode at -0.95 V vs. SCE, in aqueous tris buffer solution (pH 7), the current efficiency approached 100%. The electrochemical system did not require the presence of NADP$^+$.[14] In an alternative process, also with ICDH as the enzyme catalyst and methylviologen as the electron mediator, but using powdered CdS as a photocatalyst, the photoassisted fixation of CO_2 into oxoglutaric acid to form citric acid was accomplished, using light of $\lambda > 390$ nm. The quantum efficiency at 410 nm was 1.2%.[15]

The same reaction had been achieved with a photosensitized NADPH regeneration system, using $Ru(bpy)_3^{2+}$ as a photosensitizer, d,l-dithiothreitol as electron donor, and ferredoxin-NADP$^+$ reductase as an enzyme to recycle NADPH. Ferredoxin, which contains an iron-sulfur cluster at its redox effective site, here serves as electron mediator.[16] Ferredoxin was also used in bioelectrochemical CO_2 fixation, to mediate the isocitrate dehydrogenase catalyzed carboxylation of oxoglutaric acid to isocitric acid, and of the malic enzyme-catalyzed carboxylation of pyruvic acid to malic acid. On In_2O_3 electrodes in tris-HCl buffer solutions (pH 7.5) with 0.33 M NaCl, current efficiencies were more than 90%, with high selectivity.[17]

Using pyruvate dehydrogenase as an electrocatalyst, in the presence of NADPH, CO_2 was fixed into acetyl-coenzyme A, yielding pyruvic acid,

$$CH_3CO-SCoA + CO_2 + NADPH \rightarrow$$
$$CH_3CO-COO^- + HSCoA + NADP^+ \qquad (5)$$

In an H-type cell separated by a Nafion membrane, with a glassy carbon electrode, using $NaHCO_3$ as the CO_2 source, the turnover number per active site of the enzyme was 500. Optimal rate and yield of pyruvate formation was at pH 5.0, at an electrode potential of -0.95 V vs. SCE.[18]

A phenol carboxylase activity of *Pseudomonas aeroginosa* K$_{172}$ has been demonstrated, converting phenol selectively to 4-OH-benzoate,

The action of this enzyme was found to be Mn(II) and Fe(II) dependent, and to be stimulated by the presence of K$^+$ cations. By ^{14}C-tracer experiments, it was shown that the active carboxylating agent is the neutral CO_2 molecule and not the bicarbonate ion.[19]

BIOMIMETIC REACTIONS

Various inorganic or organic compounds may mimic the catalytic action of natural enzymes.

The natural photosynthesis by green plants was simulated by the photoreduction of CO_2 and water to formic acid and oxygen using chlorophyll-a as sensitizer. An *n*-pentane extract of chlorophyll-a was electroplated on a platinized Pt foil, which was then again platinized. This foil was placed in water saturated with CO_2 and illuminated for 30 min with a 1000-W xenon arc. On the basis of mass spectrometric analysis of the gas mixture above the aqueous solution, it was concluded that the products included formic acid, and that the oxygen produced was derived from the oxidation of water.[20]

A visible light-induced fixation of carbon dioxide with enolate complexes of aluminum porphyrins (Al-P) in benzene solutions in the presence of 1-methylimidazole resulted in formation of β-ketocarboxylate complexes of the aluminum porphyrins, which could be converted by CH_3OH/HCl treatment into the free β-ketocarboxylic acids. Thus, 1-phenyl-1-propanone was converted into 2-benzoylpropanoic acid.[21] This carboxylation of a carbonyl derivative to form a ketocarboxylic acid via an enolate as the reactive species mimics the assimilation of carbon dioxide in natural photosynthesis. Also, the photoinduced carboxylation of α,β-unsaturated esters with carbon dioxide in the presence of a methylaluminum porphyrin resulted in the formation of malonic acid derivatives. From *tert*-butyl methacrylate, ethylmethylmalonic acid mono-*tert*-butyl ester was obtained.[22,23]

Tetranuclear iron-sulfur clusters were tested as analogues of the Fe_4 active sites in iron-sulfur proteins, which play an important role in electron transfer in all living systems. Tetranuclear iron-sulfur clusters, such as $[Fe_4S_4(SR)]^{2-}$ (where R = C_6H_5 or $CH_2C_6H_5$) in DMF containing $(n\text{-}Bu_4N)(BF_4)$ as electrolyte mediated the electroreduction of CO_2 on mercury-pool cathodes. At -2 V (vs. SCE), HCOOH was produced at about 60% current efficiency. Other products were oxalate and CO. The presence of the iron-sulfur clusters caused a shift in the potential of carbon dioxide reduction by about 0.5 to 0.7 V in the positive direction.[24,25] Under similar conditions, but using Fe_4S_4 clusters bearing 36-membered methylene backbones (which were more stable than those containing thiolates), with a Hg-pool cathode and $(n\text{-}Bu_4)(BF_4)$ as electrolyte, a current efficiency of 40% was achieved for the reduction of CO_2 to HCOOH.[26]

Iron-sulfur clusters also serve as attractive catalysts for the synthesis of 2-oxazolidones, which are important as pharmaceuticals and as agricultural pesticides. The reactions were carried out at 25°C in acetonitrile solutions containing ethanolamine, a thiol such as 2-mercaptopyridine, a phosphine such as triphenylphosphine or tri-*n*-butylphosphine, and $(Et_4N)_2[Fe_4S_4(SPh)_4]$ — under an atmosphere of CO_2-O_2 (9:1). An example of such a preparation is[27]

$$Ph-CHOH-CH2-NH2 + CO2 + [Ph]3P$$

$$\xrightarrow{\text{Thiol/Fe4S4}} + [Ph]3PO$$

A simulation of the action of photosynthetic bacteria in carbon dioxide fixation as well as in nitrite or nitrate reduction to molecular nitrogen was performed in experiments with iron-sulfur clusters. Carbon dioxide fixation was coupled with nitrite iron reduction by controlled potential electrolysis (at -1.25 V vs. SCE), using $(Bu_4N)_2 [Fe_4S_4(SPh)_4]$ as electrocatalyst, in the presence of acetophenone as proton source and molecular sieve 3A as a dehydrating agent, in CO_2-saturated acetonitrile. Acetophenone was carboxylated to benzoyl acetate, while nitrite was reduced to molecular nitrogen,[28]

$$8PhCOCH_3 + 2NO_2^- + 8CO_2$$
$$+ 6e^- \rightarrow 8PhCOCH_2COO^- + N_2 + 4H_2O \quad (6)$$

In a route to α-keto acids, such as pyruvic acid, in dry acetonitrile solvent containing $(Bu_4N)(BF_4)$ and molecular sieve 3A as desiccant, iron-molybdenum and iron-sulfur clusters, $[Fe_6Mo_2S_8(SEt)_9]^{3-}$ and $[Fe_4S_4(SPh)_4]^{2-}$, were used to catalyze the reaction

$$RCO-SC_2H_5 + CO_2 + 2e^- \rightarrow RCO-COO^- + C_2H_5S^- \quad (7)$$

(where $R = CH_3, C_2H_5, C_6H_5$). The reaction apparently involved nucleophilic attack of CO_2 bound to the iron-sulfur cluster on the $R-C(O)-SC_2H_5$ molecule. With $[Fe_6Mo_2S_8(SEt)_9]^{3-}$ as catalyst, by controlled potential electrolysis using glassy carbon electrodes at -1.55 V vs. SCE, and starting with $CH_3CO-SC_2H_5$, the current yields for production of pyruvic and formic acids were 11 and 27%, respectively. Similarly, starting from $C_2H_5CO-SC_2H_5$ and from $C_6H_5CO-SC_2H_5$, the current yields of C_2H_5COOH and C_6H_5COOH were 49 and 13%, respectively. These clusters simulate the action of the enzyme pyruvate synthase.[29-31]

The same $[Fe_6Mo_2S_8(SEt)_9]^{3-}$ catalyst was used in CO_2-saturated acetonitrile to add the CO_2 moiety to methyl acrylate, $CH_2=CHC(O)OCH_3$. With glassy carbon as a working electrode at -1.6 to -1.7 V vs. SCE, in the presence of Bu_4NBF_4 and molecular sieve 4A, the main products were $CH_3-CH_2-C(O)OCH_3$ (58%) and $CH_3CH[C(O)-OCH_3]C(O)OCH_3$ (13%). The mechanism proposed involved two-electron reduction of the MoFeS cluster, which interacts with CO_2 and with H^+, enabling competitive nucleophilic addition of CO_2 or H^+ to methyl acrylate, followed by electrophilic attack of free CO_2 or H^+ to the olefinic carbon atoms.[32]

The carboxylation of active methylene compounds with CO_2 was found to be promoted by diphenylcarbodiimides, in a reaction which simulated the action of biotin as a cofactor in various enzymatic carboxylations. Thus, fluorene in DMSO solution containing an alkali carbonate was converted into fluorene-9-carboxylic acid,

The mechanism possibly involved a carbonate ion derivative of the carbodiimide,

$$C_6H_5-N(CO_2^-)-C(O^-)=N-C_6H_5,$$

as the carboxyl source. The yield of carboxylation of fluorene dependent on the nature of the alkali carbonate, and increased in the order Na < Li < Rb < K < Cs. With cesium carbonate, the yield of fluorene-9-carboxylic acid was more than 70%. Not only carbon dioxide but also the hydrogen carbonate anion were considered as the active carbon sources in the presence of carbodiimide. Other active methylene compounds which were similarly carboxylated included indene, indanone, phenylacetonitrile, acetophenone, 1-tetralone, and cyclohexanone.[33,34]

An even more convenient system for the carboxylation of active methylene compounds applied 18-crown-6 and potassium carbonate in dimethyl sulfoxide solution. At room temperature, with a reaction time of only 2 h, fluorene was carboxylated to 9-fluorene carboxylic acid (in 65% yield), cyclohexanone to 2-oxo-1-cyclohexane carboxylic acid (10%), acetophenone to benzoyl acetic acid (51%), and indene to 3-indene carboxylic acid (88%).[35]

A variety of bromomagnesium imido complexes were also found to act as effective carbon dioxide carriers, which are useful for the fixation of carbon dioxide and its transfer to active methylene compounds: (a) the 2-morpholino imidazolino Mg(II) complex,[36] (b) the N,N,-dicyclohexylamidinide Mg(II) complex,[37] and (c) the bromomagnesium thioureide complex,[38]

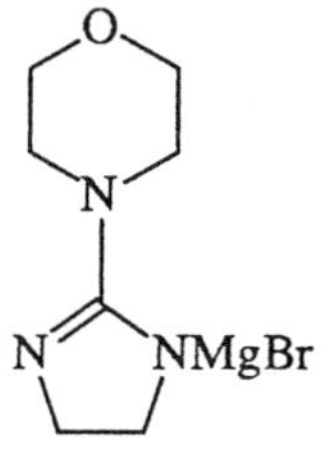

(a)

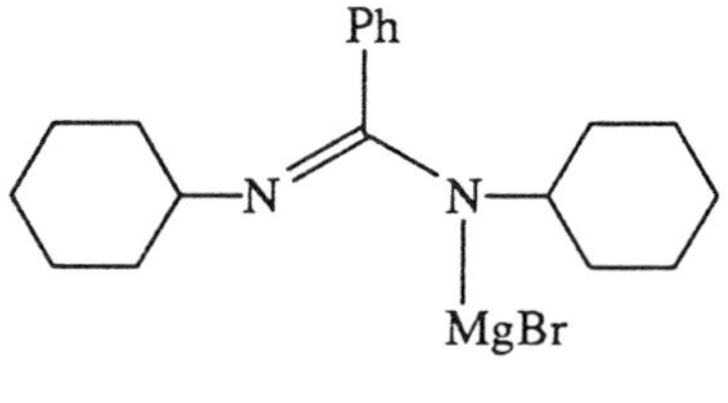

(b)

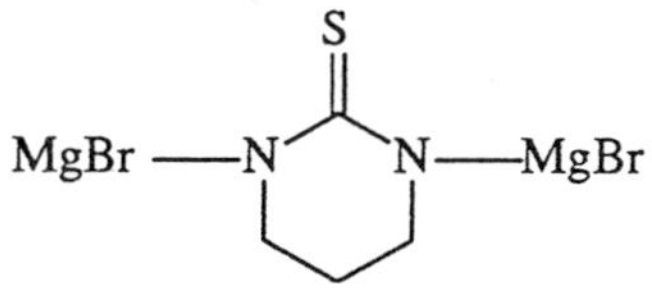

(c)

enabled the carboxylation of active methylene compounds under mild conditions, at room temperature, in THF or DMF solutions. Thus, with complex (a) and carbon dioxide, acetophenone was carboxylated to produce benzoylacetic acid, $PhCOCH_2COOH$.[36] Important applications include the carboxylation of steroids such as testosterone, androsterone, and 4-cholesten-3-one using the thioureide complex (c) in DMF solution at 15°C, forming monocarboxylated derivatives.[38]

Another interesting substrate for the fixation of carbon dioxide is cyclopentadiene. In the presence of 1,8-diazabicyclo [5.4.0] undec-7-ene (DBU) and carbon dioxide (at up to 50 kg cm^{-2} pressure) in dry DMF solution, cyclopentadiene (or its monoalkyl-substituted derivatives) was converted to dicarboxylated compounds, such as 1,3-dicarboxy cyclopentadiene,[39,40]

Enzyme-catalyzed and biomimetic reactions are extremely interesting, as they lead directly to larger molecules of considerable complexity, which may be useful and valuable, e.g., as medicines, biochemicals, vitamins, or foods.

The photoassisted reduction of carbon dioxide with suspended or colloidal semiconductors in the presence of water (see Chapter 9) may be considered a model of the fixation of CO_2 in natural plant photosynthesis (Figure 1).

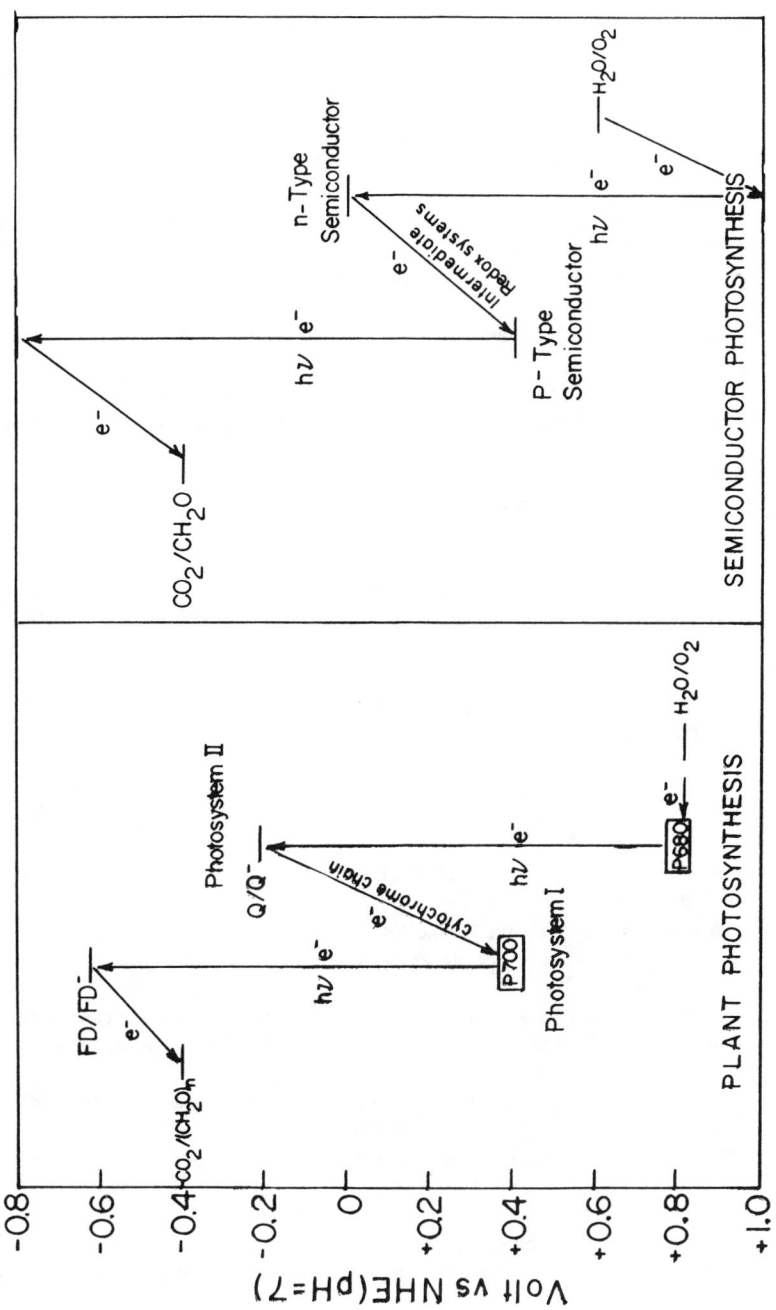

Fig. 1 Plant and semiconductor photosynthesis.

References

1. **Boardman, N. K.,** Energy from the biological conversion of solar energy, *Phil. Trans. R. Soc. London A,* 295, 477–489, 1980.

2. **Fuchs, G. and Stupperich, E.,** Carbon dioxide fixation pathways in bacteria, *Physiol. Veg.,* 21, 845–854, 1983.

3. **Fuchs, G.,** Carbon dioxide reduction by anaerobic bacteria, in *Carbon Dioxide Source Carbon: Biochemical and Chemical Uses,* (NATO ASI Ser. C), Vol. 206, 1987, 263–273.

4. **Doronina, N. V. and Trotsenko, Yu. A.,** Levels of carbon dioxide assimilation in bacteria with different pathways of C_1 metabolism, *Mikrobiologiya,* 53, 885–889, 1984.

5. **Hemming, A. and Blotevogel, K. H.,** A new pathway for carbon dioxide fixation in methanogenic bacteria, *Trends Biochem. Soc.,* 10, 198–200, 1985.

6. **Bugante, E. C., Shimonura, Y., Tanaka, T., Taniguchi, M., and Oi, S.,** Methane production from hydrogen and carbon dioxide and monoxide in a column bioreactor of thermophilic methanogens by gas recirculation, *J. Ferment. Bioeng.,* 67, 419–421, 1989.

7. **Klasson, K. T., Cowger, J. P., Ko, C. W., Vega, J. L., Clausen, E. C., and Gaddy, J. L.,** Methane production from synthesis gas using a mixed culture of *R. rubrum, M. Barkeri* and *M. formicicum, Appl. Biochem. Biotechnol.,* 24–25, 317–328, 1990.

8. **Mountfort, D. O., Pybus, V., and Wilson, R.,** Metal ion-mediated accumulation of alcohols during alkane oxidation by whole cells of *Methylosinus trichosporium, Enzyme Microb. Technol.,* 12, 343–348, 1990.

9. **Mehta, P. K., Mishra, S., and Ghose, T. K.,** Methanol biosynthesis by covalently immobilized cells of *Methylosinus trichosporium:* batch and continuous studies, *Biotechnol. Bioeng.,* 37, 551–556, 1991.

10. **Patel, R. N., Hou, C. T., and Laskin, A. I.,** Microbiological oxidation, Eur. Pat. Appl., EP 88,602 (Cl. C12N9/02), Sept. 14, 1983; *Chem. Abstr.,* 100, P33289n.

11. **Willner, I., Mandler, D., and Maidan, R.,** Bio-models and artificial models for photosynthesis, *New J. Chem.,* 11, 109–121, 1987.

12. **Mandler, D. and Willner, I.,** Photochemical fixation of carbon dioxide: enzymatic synthesis of malic, aspartic, isocitric and formic acids in artificial media, *J. Chem. Soc., Perk. Trans. II,* 1988, 997–1003.

13. **Parkinson, B. A. and Weaver, P. F.,** Photoelectrochemical pumping of enzymatic CO_2 reduction, *Nature,* 309, 148–149, 1984.

14. **Sugimura, K., Kuwabata, S., and Yoneyama, H.,** Electrochemical fixation of CO_2 in oxoglutaric acid using an enzyme as an electrocatalyst, *J. Am. Chem. Soc.,* 111, 2361–2362, 1989.

15. **Inoue, H., Kubo, Y., and Yoneyama, H.,** Photocatalytic fixation of carbon dioxide in oxoglutaric acid using isocitrate dehydrogenase and cadmium sulfide, *J. Chem. Soc. Faraday Trans.,* 87, 553–557, 1991.

16. **Willner, I., Mandler, D., and Riklin, A.,** Photoinduced carbon dioxide fixation forming malic and citric acid, *J. Chem. Soc. Chem. Commun.,* 1986, 1022–1024.

17. **Taniguchi, I.,** Electrocatalytic reduction of greenhouse gases using biofunctional metal complexes, Proc. Int. Symp. Chemical Fixation of Carbon Dioxide, Nagoya, Japan, Dec. 2–4, 1991, 81–88.

18. **Kuwabata, S., Morishita, N., and Yoneyama, H.,** Electrochemical fixation of CO_2 in acetyl-coenzyme-A to yield pyruvic acid using pyruvate dehydrogenase complexes as an electrocatalyst, *Chem. Lett.,* 1990, 1151–1154.

19. **Aresta, M., Fuchs, G., Quaranta, E., and Tommasi, I.,** Metal-ions dependence and mimetic complexes of phenol carboxylase: a new enzyme catalyzing the carboxylation of phenol using CO_2, Proc. Int. Symp. Chemical Fixation of Carbon Dioxide, Nagoya, Japan, Dec. 2–4, 1991, 353–358.

20. **Fruge, D. R., Fong, G. D., and Fong, F. K.,** Photosynthesis of polyatomic organic molecules from carbon dioxide and water by the photocatalytic action of visible-light-illuminated platinized chlorophyll *a* dihydrate polycrystals, *J. Am. Chem. Soc.,* 101, 3694–3697, 1979.

21. **Hirai, Y., Aida, T., and Inoue, S.,** Artificial photosynthesis of β-ketocarboxylic acids from carbon dioxide and ketones via enolate complexes of aluminum porphyrin, *J. Am. Chem. Soc.,* 111, 3062–3063, 1989.

22. **Komatsu, M., Aida, T., and Inoue, S.,** Novel visible-light-driven catalytic CO_2 fixation — synthesis of malonic acid derivatives from CO_2, α,β-unsaturated ester or nitrile, and diethylzinc catalyzed by aluminium porphyrins, *J. Am. Chem. Soc.,* 113, 8492–8498, 1991.

23. **Inoue, S.,** Light-induced carbon dioxide fixation mediated by metalloporphyrin, Proc. Int. Symp. Chemical Fixation of Carbon Dioxide, Nagoya, Japan, Dec. 2–4, 1991, 201–208.

24. **Tezuka, M., Yajima, T., Tsuchiya, A., Matsumoto, Y., Uchida, Y., and Hidae, M.,** Electroreduction of carbon dioxide catalyzed by iron-sulfur clusters $[Fe_4S_4(SR)_4]^{2-}$, *J. Am. Chem. Soc.,* 104, 6834–6836, 1982.

25. **Nakazawa, M., Mizobe, Y., Matsumoto, Y., Uchida, Y., Tezuka, M., and Hidai, M.,** Electrochemical reduction of carbon dioxide using iron-sulfur clusters as catalyst precursors, *Bull. Chem. Soc. Jpn.,* 59, 809–814, 1986.

26. **Tomohiro, T., Uoto, K., and Okuno, H.,** Electrochemical reduction of CO_2 catalyzed by macrocyclic Fe_4S_4 iron-sulfur clusters, *J. Chem. Soc. Chem. Commun.,* 1990, 194–195.

27. **Kodaka, M., Tomohiro, T., Lee, A. L., and Okuna, H.,** Carbon dioxide fixation forming oxazolidone coupled with a thiol/Fe_2S_4 cluster redox system, *J. Chem. Soc. Chem. Commun.,* 1989, 1479–1481.

28. **Tanaka, K., Wakita, R., and Tanaka, T.,** CO_2 fixation coupled with nitrite reduction catalyzed by $4Fe_4S$ cluster, *Chem. Lett.,* 1987, 1951–1954.

29. **Tanaka, K., Matsui, T., and Tanaka, T.,** Catalytic formation of α-keto acids by artificial CO_2 fixation, *J. Am. Chem. Soc.,* 111, 3765, 1989.

30. **Tanaka, K.,** Carbon dioxide fixation catalyzed by FeS and MoFeS clusters, Proc. Int. Symp. Chemical Fixation of Carbon Dioxide, Nagoya, Japan, Dec. 2–4, 1991, 55–62.

31. **Komeda, N., Nagao, H., Matsui, T., Adachi, G., and Tanaka, K.,** Electrochemical carbon dioxide fixation to thioesters catalyzed by $[Mo_2Fe_6S_8(SEt)_9]^{3-}$, *J. Am. Chem. Soc.,* 114, 3625–3630, 1992.

32. **Nagao, H., Miyamoto, H., and Tanaka, K.,** Carbon dioxide fixation competed with proton addition to methyl acrylate, *Chem. Lett.,* 1991, 323–326.

33. **Chiba, K., Tagaya, H., Karasu, M., Ono, T., Hashimoto, K., and Moriwaki, Y.,** The carboxylation of active methylene compounds with carbon dioxide in the presence of diphenylcarbodiimide and potassium carbonate, *Bull. Chem. Soc. Jpn.,* 64, 966–970, 1991.

34. **Chiba, K., Tagaya, H., Karasu, M., Ono, T., Saito, M., and Ashikagaya, A.,** Fixation of carbon dioxide with diphenylcarbodiimide as a model of biotin enzyme active site and a weak base: the carboxylation of fluorene under mild conditions, *Bull. Chem. Soc. Jpn.,* 64, 3738–3740, 1991.

35. **Chiba, K., Tagaya, H., Miura, S., and Karasu, M.,** The carboxylation of active methylene compounds with carbon dioxide in the presence of 18-crown-6-and potassium carbonate, *Chem. Lett.,* 1992, 923–926.

36. **Matsumura, N., Sakaguchi, Y., Ohba, T., and Inoue, H.,** 2-morpholino-imidazolino magnesium(II) complex as a novel carbon dioxide carrier, *J. Chem. Soc. Chem. Commun.,* 1980, 326–327.

37. **Matsumura, N., Ohba, T., and Inoue, H.,** The function of magnesium(II) N,N′-dicyclohexylaminide complexes as a carbon dioxide carrier, *Bull. Chem. Soc. Jpn.,* 55, 3949–3950, 1982.

38. **Matsumura, N., Asai, N., and Inoue, H.,** Fixation of carbon dioxide activated by a bromomagnesium thioureide complex, Proc. Int. Symp. Chemical Fixation of Carbon Dioxide, Nagoya, Japan, Dec. 2–4, 1991, 431–434.

39. **Haruki, E., Hara, T., and Inoue, H.,** Synthesis of tricyclo [5.2.10^{2,6}] deca-3,6-dicarboxylic acid and its derivatives from cyclopentadiene and carbon dioxide, *Chem. Express,* 5, 493–496, 1990.

40. **Haruki, E., Hara, T., and Inoue, H.,** Fixation of carbon dioxide to the cyclopentadiene derivatives using DBU and CO_2, Proc. Int. Symp. Chemical Fixation of Carbon Dioxide, Nagoya, Japan, Dec. 2–4, 1991, 427–430.

ELEVEN

Other Approaches — Outlook for Chemical Fixation

Many of the chemical reactions described in this volume are useful in their own right, for incorporating the very widely available and inexpensive carbon dioxide into important chemical intermediates, leading to polymers, pharmaceuticals, and agrochemicals. Their large-scale application thus does not depend on the intention to recycle this greenhouse gas.

ENERGY EFFICIENCY

The most immediate approach toward a decrease in the emission of carbon dioxide into the atmosphere will probably be the more efficient use of energy. It has been estimated that improved energy efficiency in the U.S. residential and commercial building sector alone (which comprises one third of the total U.S. energy consumption) could provide a savings of 50% in CO_2 emission due to this sector.[1] Comparable savings in energy consumption and hence CO_2 emission should certainly be realistic also in the industrial and transportation sectors of the economy, and achievable in all developed countries. Only after the CO_2 emission savings due to improved use of energy have been realized will it be profitable also to control the release of this greenhouse gas by chemical fixation.

If chemical CO_2 fixation is to play a role in recycling a significant part of this gas otherwise released to the atmosphere, the processes must be for high volume productions, of the order of 10^9 tons per year, or about three orders of magnitude higher than now. Hitherto the two main processes have been the production of methanol and of urea.

Primary targets for fixation of carbon dioxide will be large emission sources, such as electric power plants and steel and cement manufacture. For effective carbon dioxide recovery, it would be necessary to include CO_2 scrubbing of the waste gases from coal-fired power plants. At present, the best available technology for the recovery of carbon dioxide from central power station emissions is the absorption and stripping system using alkanolamine solvents.[2] Estimates are that such scrubbing would result in a net power loss of about 35%, and would require a plant incremental cost of 1000 to \$2000 per kWe.[3] Since the rate of release of carbon dioxide from coal- or oil-firing power plants is very large, the rate of conversion of carbon dioxide to a useful product must be equally large. If hydrogenation is to be applied, the products, such as methane or methanol, may be recycled into the power plant. The amount of carbon dioxide released during the production of hydrogen must then be considered.[4]

OCEAN DISPOSAL

One proposal for chemical fixation, albeit not to useful products, has been to pump the CO_2-rich combustion gases from fossil-fuel power stations into the ocean, say at about 90 m depth, to release it through manifolds. The dissolved magnesium and calcium salts in the naturally slightly alkaline oceanwater (pH 8.0) would cause the precipitation of the carbon dioxide as carbonates, thus preventing its release to the atmosphere. The process would be possible only if tidal or other currents cause quite rapid replacement of the water mass around the exhaust manifold, to renew the required magnesium and calcium salts. This inorganic "sink" for carbon dioxide would be in addition to assimilation by phytoplankton.[5] Other proposals are to enhance the growth of oceanic biomass, e.g., by cultivating the naturally growing macroalgae, such as *Macrocystis, Laminaria, Gracilaria,* and *Euchema,* which form large kelp beds on the continental shelves. These can be harvested, providing useful chemicals and foods, and can be converted to synthetic natural gas. A more controversial proposal is to stimulate the growth of microalgae in the open ocean by fertilization with a limiting nutrient, such as iron salts.[3]

An alternative approach to chemical fixation could be to remove the carbon dioxide contained in the flue gases of large electricity generating stations and cool and compress it to liquid carbon dioxide. It could then be released in the deep ocean, i.e., more than 1000 m depth, where it would be sequestered in clathrate form. The theoretical equation for the clathrate formation is given by[6]

$$CO_2 + (5.75)H_2O = CO_2.(5.75)H_2O \quad \Delta H = 14.43 \text{kcal/mol (at 276.9 K)}$$

The crystalline CO_2 clathrate was found to have a unit cell consisting of 46 water molecules. These enclose six large holes and two small holes, which can encage carbon dioxide molecules. At the deep-sea temperature of about 275 K, the CO_2 clathrate is formed at a pressure higher than 1.18 MPa. Because of the higher density of the carbon dioxide clathrate (1.1 g cm^{-3}) compared with the density of seawater (1.07 g cm^{-3} at 1000 m depth), it would presumably sink to the ocean floor.[6] However, it has been estimated that the removal of 50 or 90% of the CO_2 in this way would either double or triple the cost of electricity.[3,7-9] Also, the long-term stability and the environmental consequences of large quantities of CO_2 in the ocean are yet unknown.[2]

In order to avoid the high costs of deep ocean injection of carbon dioxide, another proposal has been to inject the gas into the sea at a shallow depth of several hundred meters on the continental shelf. The negative buoyancy of CO_2-enriched seawater would pull the gas down from the emission sites into the deep ocean. The problems to be considered would be the effect of the expected pH decrease of the seawater, to about pH 4 to 5, which probably would have serious adverse effects on marine life.[10]

SUPERCRITICAL CARBON DIOXIDE

Not enough research has been devoted to the use of supercritical carbon dioxide as a reaction medium. This promises to be a safe and environmentally acceptable alternative to halocarbon and hydrocarbon solvents. Supercritical-fluid extraction (SFE) enables extraction of heavy nonvolatile substances in supercritical fluids. The high solubility is possible because of the high density of the supercritical fluid, which is close to that of the liquid. For pure CO_2, $T_c = 304$ K, $P_c = 7.38$ MPA (1070 psi), and $\rho_c = 0.468$ g cm^{-3}. SFE processes with carbon dioxide have until now been particularly favored for the food and pharmaceutical industries, as CO_2 is nontoxic (in contrast to halocarbon solvents such as methylene chloride or chloroform which are carcinogenic) and noninflammable (in contrast to hydrocarbons), and leaves no harmful residues. The major application has been for caffeine extraction from coffee beans.[11]

Novel uses of supercritical CO_2 are gradually appearing, such as for chemical reactions. Liquid CO_2 was used to prepare to bind carbon dioxide into transition-metal complexes. Thus, the complex $[Ni(PCy_3)_2]_2(\mu\text{-}N_2)$, when placed in liquid CO_2, was converted in high yield into $[Ni(PCy_3)_2]_2(\mu\text{-}CO_2)$. Similar reactions occurred with some iridium and cobalt complexes.[12] The Diels-Alder reaction between isoprene and methyl acrylate in supercritical carbon dioxide was studied, using a stainless steel reactor, at a temperature of 323 K and pressures in the range of 4.9 to 20.6 MPa. The mechanism of the reaction was investigated by FTIR spectroscopy. After the reaction, the cell effluent was analyzed by gas chromatography. In supercritical CO_2 there occurred considerable aggregation and cluster formation by solvent molecules about the activated complex of the Diels-Alder reaction, since the compressibility of the solvent is enhanced in the near-critical region. The product isomer distribution by reaction in the critical region was different from that performed at atmospheric pressure. This was explained to be due to the aggregation of the solvent molecules, causing steric hindrance to the reactants.[13]

An interesting application of supercritical carbon dioxide is in the synthesis of fluoropolymers. These polymers are very valuable, e.g., as high performance lubricants in computer disk drives, as aircraft fuel sealants, and in protective coatings. Fluoropolymers are insoluble in most common solvents, except for chlorofluorocarbons (CFCs) — which are undesirable because of their contribution both to the greenhouse effect and to the destruction of the upper atmosphere ozone layer. Highly fluorinated acrylic monomers may be polymerized in homogeneous solution in supercritical carbon dioxide. 1,1-Dihydrofluorooctyl acrylate (FOA) was polymerized in supercritical CO_2, at up to 207 bar pressure and 59.4°C, using azo bis isobutyronitrile (AIBN) as initiator. Even high molecular mass (>250,000 g mol^{-1}) homopolymers of various acrylic monomers are soluble in liquid CO_2, up to 25% (w/v). By contrast, hydrocarbon polymers, such as methyl methacrylate, are practically insoluble in liquid CO_2. Presumably, there exists a specific solute-solvent interaction between the CO_2 solvent and the fluorocarbon. Supercritical CO_2 was also a satisfactory solvent for the co-polymerization between fluorinated acrylic monomers and conventional hydrocarbon monomers, such as methyl methacrylate, butyl acrylate, styrene, and ethylene.[14]

Supercritical carbon dioxide is also slowly finding special uses, e.g., as a means of aerosol or vapor generation, such as for the introduction of samples of organic solutions of metal complexes, in atomic absorption spectrometry.[15]

CONCLUSIONS

The more hopeful results of the effort of the last decade of CO_2 fixation research have been in

1. The application of gas-diffusion electrodes, permitting high current densities for the electrochemical production of small molecules.
2. The introduction of enzyme-catalyzed electrochemical and photochemical reactions, leading to valuable biochemicals.

If sufficient research effort will be provided, the target of halting the increase in atmospheric carbon dioxide may yet be achieved!

References

1. **Rosenfeld, A. H. and Price, L.,** Options for reducing carbon dioxide emissions, *Global Warming: Physics and Facts, AIP Conf. Proc.,* 247, 261, 1992.
2. **Steinberg, M., Lee, J., and Morris, S.,** An assessment of CO_2 greenhouse gas mitigation technologies, BNL 46045, Brookhaven National Laboratory, Upton, NY, 1991.
3. **Spencer, D. F.,** A preliminary assessment of carbon dioxide mitigation options, *Annu. Rev. Energy Environ.,* 16, 259–273, 1991.
4. **Mizuno, K. and Misono, M.,** Assessment of catalytic technology for the reduction of CO_2 and other greenhouse gases, Proc. Int. Symp. on Chemical Fixation of Carbon Dioxide, Nagoya, Japan, Dec. 2–4, 1991, 237–242.

5. **Wisseroth, K.,** Problem of atmospheric carbon dioxide and its possible control by an ocean system, *Chemiker-Zeitung,* 115, 45–52, 1991.

6. **Saji, A., Yoshida, M., Sakai, M., Tanii, T., Kamata, T., and Kitamura, H.,** Fixation of carbon dioxide by clathrate-hydrate, Proc. Int. Symp. Chemical Fixation of Carbon Dioxide, Nagoya, Japan, Dec. 2–4, 1991, 443–448.

7. **Steinberg, M.,** An analysis of concepts for controlling atmospheric carbon dioxide, BNL-33960, Brookhaven National Laboratory, Upton, NY, 1983.

8. **Steinberg, M. and Cheng, H. C.,** A systems study for the removal, recovery, and disposal of carbon dioxide from fossil fuel power plants in the U.S., BNL-35666, Brookhaven National Laboratory, Upton, NY, 1985.

9. **Mintzer, I. M.,** Energy, greenhouse gases, and climate change, in *Annual Reviews of Energy,* Vol. 15, Hollander, J. M., Socolow, R. H., Sternlicht, D., Eds., 1990, 513–538.

10. **Haughan, P. M. and Drange, H.,** Sequestration of CO_2 in the deep ocean by shallow injection, *Nature,* 357, 318–320, 1992.

11. **Johnston, K.,** Supercritical fluids, *Kirk-Othmer Encyclopedia of Chemical Technology,* 3rd ed., (suppl. vol.,) 1984, 872–893.

12. **Mason, M. G. and Ibers, J. A.,** Reactivity of some transition-metal systems toward liquid carbon dioxide, *J. Am. Chem. Soc.,* 104, 5153–5157, 1982.

13. **Ikushima, Y., Saito, N., and Arai, M.,** Supercritical carbon dioxide as reaction medium: examination of its solvent effect in the near-critical region, *J. Phys. Chem.,* 96, 2293–2297, 1992.

14. **Desimone, J. M., Guan, Z., and Elsbernd, C. S.,** Synthesis of fluoropolymers in supercritical carbon dioxide, *Science,* 257, 945–947, 1992.

15. **Bysouth, S. R. and Tyson, J. F.,** Supercritical carbon dioxide as a carrier for sample introduction in atomic absorption spectrometry, *Anal. Chim. Acta,* 258, 55–60, 1992.

INDEX

A

Absorption, 39, 69, 132
Acenaphthylene, 91
<*trans*-Acenophthene-1,2-dicarboxylic acid, 91
Acetaldehyde, 136
Acetic acid, 143
Acetone, 34
Acetonitrile, 13, 23, 58, 88, 92, 98
Acetophenone, 148, 149
Acetyl-CoA, 143
Acetylene, 47
Acetylsalicylic acid, 20
Acid-base equilibrium, 58
Acid rain, 1
Acrylic acids, 92
Activation energy, 12, 49
Adsorption, 70, 79
    ammonia, 7
    capacity for, 39
    carbon dioxide, 7
    co-, 7
    dissociative, 46–47
    energy of, 76
    hydrogen, 69
    molecular, 6
AES, see Auger electron spectroscopy
AIBN, see Azo bis isobutyronitrile
Alanine, 62
Alcohols, see also specific types
    distribution of, 77
    electrochemical reduction to, 78–81
    photochemical reduction and, 62
    propargyl, 19
    synthesis of, 44
Aldehydes, 77, 91, see also specific types
Aliphatic amines, 17, see also specific types
Alkaline carbonates, 40, see also specific types
Alkaline fuel cells, 100
Alkalines, 40, see also specific types
Alkali promoters, 6
Alkanes, 36, 77, see also specific types
Alkanolamine solvents, 156
Alkenes, 20, 77, see also specific types
N-Alkylcarbamate esters, 17

Alkynes, 18, 19, 20, 91, see also specific types
Alloy electrodes, 81–82
Alumina, 37, 39
β-Alumina solid electrolytes, 105
Aluminum acetate, 17
Aluminum anodes, 89, 90
Aluminum-carbamate, 17
Aluminum oxalate, 89
Aluminum porphyrins, 17, 147
Amides, 45, see also specific types
Amines, 15, 16, see also specific types
    aliphatic, 17
    conversion of to amino acids, 62
    photochemical reduction and, 61, 62
    primary, 17
    secondary, 17, 19, 24, 62
    tertiary, 19, 58
Amino acids, 62, see also specific types
α-Aminonitriles, 19
Ammonia, 7, 15, 37
Ammonium carbamates, 7, 15, 18
Ammonium cyanate, 15
Ammonium tetrafluoroborate, 96
Amorphous carbon, 102
Androsterone, 150
Aniline, 125
Antiinflammatory agents, 91, see also specific types
Antitumor agents, 26, see also specific types
Aqueous solutions, see also specific types
    electrochemical reduction in, see under Electrochemical reduction
    heterogeneous photo-assisted reduction in, 131–139
Argon, 96
Aromatic compounds, 61, see also specific types
Aromatic hydrocarbons, 40, 60, see also specific types
Aromatization, 36
Arrhenius plots, 49
Artificial carboxylation, 145
Artificial photosynthesis, 5, 57
α-Arylpropionic acids, 91
Ascorbate buffer, 60
Aspartic acid, 145
Aspirin, 20

Transition metals, 23, 40, 45, 82, 95, see also
specific types
Transmission electron microscopy (TEM), 49
Triethanolamine, 16, 19, 138
Triethylamine, 25, 59, 60
Triphenylphosphine, 26
Trisodium phosphate, 7, 8
Tungsten carbides, 100

U

Ultrasound, 48
Ultraviolet illumination, 57–58
β-Unsaturated acids, 91
Uranyl ions, 58
Ureas, 15, 18, 19
Uron herbicides, 18, see also specific types

V

Vanadium redox couple, 125
Vinyl carbamates, 19
Voltammetry, 69, 96, 100
Voltammograms, 83

W

Waste carbon dioxide, 16

Water gas shift reaction, 8, 33, 36, 38, 41,
46–47
Wind energy, 3
Woehler's preparation of urea, 15
Wustite, 37

X

XPS, see X-ray photoelectron spectroscopy
X-ray analysis, 26, see also specific types
X-ray diffraction (XRD) analysis, 105, 140
X-ray photoelectron spectroscopy (XPS), 7, 77
X-ray photon spectroscopy, 77
XRD, see X-ray diffraction

Y

YSZ, see Ytrrium-stabilized zirconia
Yttrium-stabilized zirconia (YSZ), 103

Z

Zinc, 36, 40, 71, 136
Zinc anodes, 89
Zinc oxalate, 89
Zinc oxide catalysis, 35
Zinc porphyrins, 62
Zinc sulfide, 137
Zirconia, 6, 41, 103, 104
Zirconium, 41

$$CO_2 \quad H_2 \qquad C_2H_5OH \qquad H_2O$$

$$\left(\; 2(CO_2) + 6H_2 = C_2\underline{H_5}OH + \underline{3H_2O} \right)$$

C	2		C	2
H	12		H	12
O	4		O	4

$$5,420 \quad cal/gm$$

rsc

rcratey

estec04

autothanol methend refames

PCCP 3 2001